Isaac Physics Skills

Essential GCSE Physics

A.C. Machacek & K.O. Dalby
Westcliff High School for Boys

with extra questions written by R. Meikle

Periphyseos Press
Cambridge, UK.

Co-published in Cambridge, United Kingdom, by
Periphyseos Press and Cambridge University Press.

`www.periphyseos.org.uk` and `www.cambridge.org`

Essential GCSE Physics

First published, 1st & 2nd reprints 2017
3rd & 4th reprints 2018; 5th reprint 2019
6th reprint 2020
Co-published adaptation, 2020

Printed and bound in the UK by Short Run Press Limited, Exeter.

Typeset in LATEX

A catalogue record for this publication is available from the British Library

ISBN 978-1-8382160-1-6 Paperback

Use this collection of worksheets in parallel with the electronic version at
`isaacphysics.org`. Marking of answers and compilation of results is free
on Isaac Physics. Register as a student or as a teacher to gain full
functionality and support.

 used with kind permission of M. J. Rutter.

Notes for the Student and the Teacher

Some students (and teachers) may seem daunted by the idea of performing calculations on momentum conservation, lens images, circular motion, nuclear energy and the like, below the age of 16. However, it is our conviction that all secondary school students can gain mastery of these concepts. By mastery, we mean that you feel confident that you understand them and can apply them to reasonably straightforward problems with accuracy. To begin, all you need is the ability to multiply and divide numbers (using a calculator), the willingness to give Physics a go, and the determination not to listen to any thoughts which say that it's going to be too difficult.

Each worksheet contains notes, explanations, worked examples and then questions. It is important that you can see where formulae come from, and accordingly explanations in places will go deeper than required for GCSE.

Once you have read the notes, you are ready to try the questions, they get harder as you go on. You can check your answers using the isaacphysics.org website.

We suggest that you revisit each sheet and its questions until you can answer at least three quarters of the material correctly; see the pass mark indicated in the square on each sheet. Until 75% is achieved, study further, then repeat a selection of questions. This is the mastery method here ensuring a good foundation is laid for a GCSE physics education.

$$^x/_y$$

You may well also find this book helpful when you come to revise. When you revise, resist the temptation to move beyond a page until you have attempted a good selection of the questions and have got at least three-quarters of them correct. Be aware that this book is not written with a particular specification in mind - not all sections will be relevant to your exam, especially those marked $\heartsuit$. A specification map is on isaacphysics.org. $\heartsuit$

Also remember, this is a Physics book not an exam revision guide. Its aim is to help you understand the principles. This may well take longer than memorizing a few soundbites, but, once achieved, enables you to solve a wide range of problems with very little further 'learning'. Quite simply, as thousands of our students have found, once you 'get it' it is then 'obvious' and no further notes or books are needed. This is mastery - and this is what you should aim for.

ACM & KOD
Westcliff-on-Sea, 2016

Acknowledgements

We are very grateful to the Isaac Physics team at the University of Cambridge for initiating this project, and for their continued advice, flexibility and encouragement. Particular thanks go to Aleksandr Bowkis, Ben Hanson, Rupert Fynn and Umberto Lupo for their great work in typesetting our sheets into a book. Michael Conterio and Bianca Andrei have closely checked the questions while making them available through the isaacphysics.org website, a remarkable resource for students and teachers alike. Laura Moat must be thanked for all of her expertise and help with the aesthetics of these worksheets, both on the covers and the content itself.

We must also thank Rob Meikle for his generosity in allowing us to use many of his excellent Physics questions from his rich resource 'Physics Examples' within many of the worksheets.

We are also grateful to Jennifer Crowter and the Physics Department of the Royal Grammar School, High Wycombe, where some early versions of the exercises were tested, and whose ongoing support has been helpful.

We also thank our colleagues at Westcliff High School for Boys. Joy Williams, Wayne Williams, Kerrie Mumford, Freyja Dolan, Tim Sinnott and Iain Williamson (not to mention our students) gave helpful thought to the educational philosophy of the project. Simon Hudson and Harry Tresidder helped us test material in the classroom. Furthermore, we particularly appreciate the strong support and encouragement given to us by Mr Michael Skelly, the Headmaster, who has not only been willing for his School to be used as a testing ground, but enthusiastically gave us time to collaborate with Cambridge on this project.

Still deeper thanks are owed to Steph Dalby and Helen Machacek for their ongoing love, encouragement, advice and support – not to mention their willingness for us to be involved in this project despite the reduction in the time we then had available for our families. Steph's experience of using this material in non-selective school settings, and regular contact with science teachers who are not specialist physicists has given an invaluable perspective to our work; as has Helen's detailed knowledge of mathematics in the primary school. We are extremely fortunate to have such partners on the journey of life.

Finally, we thank you, the student. We wish you well in your studies, and trust that with a bit of determination, you will use these to help you become a master of physics at GCSE level, and that this might encourage you to try more physics in the sixth form.

Soli Deo Gloria,

ACM & KOD
Westcliff-on-Sea, 2016

Suggestions for use in lessons

Traditional approach

- Introduce the concept you wish to teach – perhaps by giving an example of a situation where this is going to be useful in the solution. A problem could be set, or a short video of a relevant situation shown (I like showing a YouTube clip of a North American 'Fall' festival in which a 500 kg pumpkin is dropped on an old school bus, when about to discuss gravitational and kinetic energy, or the Mythbusters video in which a compressed gas cylinder bursts through a breeze block wall when its regulator is sheared off, as an introduction to gas pressure).

- If desired, you can project the relevant page from the teacher's pdf – with essential details, but with some key words and definitions missing, and spaces for certain explanations. Before students have opened their Isaac books, teach the main concepts, give the definitions and use class questioning and discussion to agree the answers to the 'cloze text' parts: isaacphysics.org/books/phys_book_gcse

- Students then turn to the relevant page, and read the 'notes' section. Students may make their own notes in their exercise books if you wish.

- While many students will be ready to begin the questions straight away (and will be able to do so without further help from you), others will need your specific help in going over important points in the notes.

- By the time that the students who needed your help with the notes are ready to start the more straightforward questions, the others will have reached the more tricky questions, and will wish assistance. Questions can be answered in students' exercise books, showing working, with the final answer to an appropriate number of significant figures and with a unit. You may choose to make certain questions optional.

- Follow-up questions can be selected from the Isaac Physics website for homework.

Use with 'Flipped Lessons'

- Here, you would set a homework to study the notes of a particular page, and complete some of the more straightforward questions – you can check their progress using isaacphysics.org.

- Students then work on questions in class, as above. The teachers' version of the text (with spaces for the explanations and results of class discussion) could be projected onto the screen and discussed as the starter for the lesson to see how much students remember and understand from their own reading.

Using Isaac Physics with this book

Isaac Physics offers online versions of each sheet at:

isaacphysics.org/gcsebook

There, a student can enter answers as well as learn the concepts detailed in these worksheets by reading the online versions. This online tool will mark answers, giving immediate feedback to a student who, if registered on isaacphysics.org, can have their progress stored and even retrieved for their CV! Teachers can set a sheet for class homework as the appropriate theme is being taught, and again for pre-exam revision. Isaac Physics can return the fully assembled and analysed marks to the teacher, if registered for this free service. Isaac Physics aims to follow the significant figures (sf) rules on page (v), and warns if your answer has a sf problem. Isaac is stricter at A-level, in accordance with A-level examination practice.

Uncertainty and Significant Figures

In physics, numbers represent measurements that have uncertainty and this is indicated by the number of significant figures in an answer.

Significant figures
When there is a decimal point (dp), all digits are significant, except leading (leftmost) zeros: 2.00 (3 sf); 0.020 (2 sf); 200.1 (4 sf); 200.010 (6 sf)
Numbers without a dp can have an *absolute accuracy*: 4 people; 3 electrons. Some numbers can be ambiguous: 200 could be 1, 2 or 3 sf (see below). Assume such numbers have the same number of sf as other numbers in the question.

Combining quantities
Multiplying or dividing numbers gives a result with a number of sf equal to that of the number with the smallest number of sf:
$x = 2.31, y = 4.921$ gives $xy = 11.4$ (3 sf, the same as x).
An absolutely accurate number multiplied in does not influence the above.

Standard form
Online, and sometimes in texts, one uses a letter 'x' in place of a times sign and ∧ denotes "to the power of":
1 800 000 could be 1.80x10^6 (3 sf) and 0.000 015 5 is 1.55x10^−5
(standardly, 1.80×10^6 and 1.55×10^{-5})
The letter 'e' can denote "times 10 to the power of": 1.80e6 and 1.55e−5.

Significant figures in standard form
Standard form eliminates ambiguity: In $n.nnn \times 10^n$, the numbers before and after the decimal point are significant:
$191 = 1.91 \times 10^2$ (3 sf); 191 is $190 = 1.9 \times 10^2$ (2 sf); 191 is $200 = 2 \times 10^2$ (1 sf).

Answers to questions
In these worksheets and online, give the appropriate number of sf:
For example, when the least accurate data in a question is given to 3 significant figures, then the answer should be given to three significant figures; see above.
Too many sf are meaningless; giving too few discards information. Exam boards require consistency in sf, so it is best to get accustomed to proper practices.

Contents

Skills

In Physics, measurable quantities usually have a number and a unit. The unit gives an indication of the size of that quantity and also information about what the quantity physically represents. This is best understood with examples.

A quantity such as 15 metres is clearly a length; one cannot measure a mass or a time in metres. 15 metres is a shorter length than 15 miles, but a longer length than 15 inches. Without the inclusion of a unit, a length of 15 is meaningless.

To facilitate global collaboration in science, seven units have been selected as the standard that all scientists should use. These are called SI base units (which comes from the French name: Système International d'unités). At GCSE Physics level, you are expected to know and be able to use the first six of these units.

Quantity	Unit name	Unit symbol
Length	metre	m
Mass	kilogram	kg
Time	second	s
Electric current	ampere	A
Temperature	kelvin	K
Amount of substance	mole	mol
Luminous intensity	candela	cd

SI derived units are units given in terms of the SI base units. A speed, for example, is always a length divided by a time. In SI derived units, a speed should be given in metres per second (m/s). A volume always includes the product of three lengths so, in SI derived units, a volume should be given in cubic metres (m^3).

You can work out what the appropriate unit for any quantity is by considering the quantities that are combined in any equation for that quantity.

Units may also include a prefix. These are included between the number and the unit and tell you by how much the number should be multiplied.

Prefix	Multiply By
mega (M)	1 000 000
kilo (k)	1 000
centi (c)	0.01
milli (m)	0.001
micro (µ)	0.000 001
nano (n)	0.000 000 001

1.1   Complete the table below with the correct SI base units.

Quantity	Equation	Unit in terms of SI base units
Area	$A = L^2$	(a)
Acceleration	$a = (v - u)/t$	(b)
Momentum	$p = mv$	(c)
Kinetic energy	$E = \frac{1}{2} mv^2$	(d)
Gravitational potential energy	$E = mgh$	(e)
Electric charge	$Q = It$	(f)

1.2   Write the following quantities with the appropriate unit and prefix

0.000 001 20 m	(a)	5 200 000 mg	(b)
6 500 µs	(c)	0.000 000 920 km	(d)
3 400 000 nA	(e)	0.000 027 0 kA	(f)
5 500 000 000 nm	(g)	6 500 000 cm²	(h)
0.000 044 0 km/s	(i)	83 000 mm³	(j)

1.3   Convert these measurements to metres (m):

(a) 240 cm   (b) 1 500 cm   (c) 95 cm   (d) 7.0×10^3 cm

1.4   Convert these mass measurements into kilograms (kg):

    (a) $2\,500$ g   (b) 350 g   (c) $1\,020$ g   (d) 3.80×10^4 g

1.5   Convert these mass measurements into grams (g):

    (a) 6.70 kg   (b) $3\,400$ mg   (c) 0.050 kg   (d) 150 mg

1.6   Convert the following volumes into cubic metres (m^3) [$1\,cm^3 = 1\,ml$]:

    (a) $2\,500\,cm^3$ (b) $68\,cm^3$ (c) $3\,700$ litres

1.7   Convert the following volumes to litres (L):

    (a) $2\,500\,cm^3$ (b) $2.0\,m^3$ (c) $560\,cm^3$

1.8   How many cubic centimetres (cm^3) are there in these volumes?

    (a) 1.60 litres   (b) $3.25\,m^3$ (c) $0.0625\,m^3$ (d) 0.080 litres

1.9   Convert these areas into square metres (m^2):

    (a) $4\,250\,cm^2$ (b) $5.3 \times 10^4\,cm^2$ (c) $2.50\,km^2$ (d) $15.0\,cm^2$

1.10   Calculate the number of square centimetres (cm^2) in:

    (a) $1.44\,m^2$ (b) $0.0275\,m^2$ (c) $3.50 \times 10^{-2}\,m^2$ (d) $1.50 \times 10^{-4}\,m^2$

$^{35}/_{46}$

Additional Units Questions

1.11   Change these times into seconds (s):

    (a) 3.0 mins   (b) 2 hrs 30 mins   (c) 3.6 mins   (d) 4 mins 30 secs

1.12   How many seconds are there in a minute, an hour, a day and a year?

1.13   Write the following fundamental constants and data without unit prefixes.

    (a) speed of light $= 300$ Mm/s   (b) $g = 9\,810$ mN/kg

    (c) Earth's radius $= 6\,370$ km   (d) red wavelength $= 680$ nm

1.14   The light-year (ly) is a unit often mistaken as a unit of time. It is defined as the distance travelled by light in a vacuum in one Julian year (365.25 days). Use the data in Q1.13 and the equation speed $=$ distance/time ($v = s/t$). What SI measurement is 1.0 ly equivalent to?

2   Standard Form

The radius of the Earth is 6 400 000 m.
The speed of light is 300 000 000 m/s.
The charge of one electron is $-0.000\,000\,000\,000\,000\,000\,16$ C.

Big and small numbers are inconvenient to write down – scientists and engineers use standard form to make things clearer.

The above numbers in standard form look like this:

$$6.4 \times 10^6 \text{ (or 6.4 e 6 on a computer).}$$

$$3.0 \times 10^8 \text{ (or 3.0 e 8 on a computer).}$$

$$-1.6 \times 10^{-19} \text{ (or } -1.6\,\text{e}-19 \text{ on a computer).}$$

number in standard form = mantissa × power of ten

The mantissa is a number bigger than or equal to 1, but less than 10.

2.1   Which of the following numbers could be a mantissa?
(a) 9.5 – *Yes, it is larger than (or equal to)* 1 *and smaller than* 10
(b) 0.4 – *No, it is smaller than* 1
(c) 12.3 – *No, it is not less than* 10
(d) 0.2   (h) 0.04
(e) 1.2   (i) 10
(f) 1.0   (j) 5
(g) 10.3   (k) 7.6

Powers of ten are numbers you can make by starting with 1 and either multiplying or dividing as many times as you like by 10.

So 100, 0.01, 100 000, 10, 1 and 0.000 1 are all powers of ten,
but 30, 0.98 and 40 000 are not powers of ten.

Powers of ten can be written using exponents (e.g. 10^2 rather than 100).

$$10\,000 = 10 \times 10 \times 10 \times 10 = 10^4 \quad \text{exponent} = 4$$
$$1\,000 = 10 \times 10 \times 10 = 10^3 \quad \text{exponent} = 3$$
$$100 = 10 \times 10 = 10^2 \quad \text{exponent} = 2$$
$$10 = 10 = 10^1 \quad \text{exponent} = 1$$
$$1 = 10^0 \quad \text{exponent} = 0$$
$$0.1 = 1/10 = 1/10^1 = 10^{-1} \quad \text{exponent} = -1$$
$$0.01 = 1/100 = 1/10^2 = 10^{-2} \quad \text{exponent} = -2$$
$$0.001 = 1/1000 = 1/10^3 = 10^{-3} \quad \text{exponent} = -3$$

2.2 For the following numbers, decide if they are powers of ten and, if they are, write down the exponent. The first two have been done for you.

Number	Power of Ten	Exponent
30	✗	
100	✓	2
0.004	(a)	
0.01	(b)	
1 000 000 000	(c)	
0.000 000 1	(d)	

Note that the exponent counts the number of times the decimal point must be moved to get from its starting point before the number is turned into a mantissa:

e.g. $0.\,0\,0\,0\,3\,2 = 3.2 \times 10^{-4}$, as the decimal point must be moved 4 times to the right before it makes a 3.2.

Also $893 = 8\,9\,3\,.0 = 8.93 \times 10^2$ as the decimal point must be moved 2 times to the left before it makes an 8.93.

2.3 For each of the following numbers state how many times the decimal point must be moved ($+$ve to the left, $-$ve to the right) when making the numbers into a mantissa and the exponent of the 10 when in standard form.

(a) 0.000 145   (c) 345 094   (e) 69 023   (g) 0.011 2

(b) 153.034 2   (d) 0.003 425 39   (f) 0.000 002 87   (h) 56 920.142 2

2.4   Write the following numbers as mantissa × power of ten, then write the exponent, and finally write them in standard form. The first two are done as examples.

Number	Mantissa × Power of 10	Standard Form
450 000	$4.5 \times 100\,000$	4.5×10^5
0.000 032	$3.2 \times 0.000\,01$	3.2×10^{-5}
300	(a)	(b)
0.026	(c)	(d)
390 000	(e)	(f)
6 700	(g)	(h)
0.000 000 062	(i)	(j)

2.5   Express the following numbers in standard form:
(a) 4 000   (c) 8.31   (e) 860 000   (g) 920   (i) 435 981 719
(b) 0.030   (d) 0.000 002 8 (f) 0.002 451 (h) 0.109 3 (j) 0.000 004 72

2.6   Write the following numbers in the normal way (e.g 3 300):
(a) 3×10^3 (d) 76×10^{-3} (g) 5.23×10^{-7} (j) 3.5×10^{-2}
(b) 2×10^{-2} (e) 3.54×10^0 (h) 3.2185×10^{-4} (k) 8.54×10^7
(c) 6×10^1 (f) 9.73×10^8 (i) 6.9836×10^5 (l) 1.25×10^{-1}

You key 3.4×10^{-9} into a calculator by pressing

| 3 | . | 4 | $\times 10^n$ | – | 9 |

2.7   Do the following calculations on your calculator.
(a) $(3.0 \times 10^8) \div (6.6 \times 10^{-7})$
(b) $(3.0 \times 10^8) \div (3 \times 10^{-2})$
(c) $(3.0 \times 10^8) \div (2.3 \times 10^2)$
(d) $(3.0 \times 10^8) \div (5 \times 10^{-11})$

3 Rearranging Equations

Whatever is done to one side of an equals sign must be done to the other also. Take, for example, the equation:

$$a = b + c$$

a is the subject. To make b the subject, one must look at what is done to b and do the inverse to both sides. In the above equation, c is added to b, so b is made the subject by subtracting c from both sides of the equals sign:

- Subtracting c: $a - c = b + c - c$

- Simplifying the right hand side: $a - c = b$

- Writing b as the subject: $b = a - c$

Addition and subtraction are inverse operations.

Multiplication and division are inverse operations.

Powers and roots are inverse operations.

Example 1 – Make y the subject of $x = 2 \times y + z$

The last operation on y is the addition of z, so subtract z from both sides:

$$x - z = 2 \times y$$

y is multiplied by 2, so divide both sides of the equation by 2:

$$(x - z)/2 = y$$

Example 2 – Make g the subject of $5\sqrt{g} = h + j$

Divide by 5:

$$\sqrt{g} = (h + j)/5$$

Square both sides:

$$g = (h + j)^2/25$$

3.1   Rearrange the following equations to make the variable in brackets the subject:

(a) $p = mv$ (m) (f) $M = Fd$ (d)

(b) $Q = It$ (I) (g) $V/R = I$ (R)

(c) $v = s/t$ (s) (h) $P/I = V$ (P)

(d) $F = ma$ (a) (i) $v = f\lambda$ (λ)

(e) $W = mg$ (m) (j) $\rho = m/V$ (V)

3.2   Rearrange the following equations to make the variable in brackets the subject:

(a) $E = mgh$ (m)

(b) $P_1 V_1 = P_2 V_2$ (P_2)

(c) $v^2 = u^2 + 2as$ (a)

(d) $\sin(c) = 1/n$ (n)

(e) $V_p/V_s = N_p/N_s$ (N_s)

3.3   Make v the subject of the following equation:
$$E = \tfrac{1}{2}mv^2$$

3.4   If $u = 0$, make t the subject of the following equation:
$$s = ut + \tfrac{1}{2}at^2$$

3.5   Make $\sin(r)$ the subject of the following equation:
$$n = \frac{\sin(i)}{\sin(r)}$$

3.6   Make x the subject of the following equation:
$$10(x + y) = 5(x - y)$$

3.7   Make λ the subject of the following equation:
$$t = k/\lambda$$

3.8   Make r the subject of the following equation:
$$F = \frac{kQ_1Q_2}{r^2}$$

3.9   Make T the subject of the following equation:
$$r\left(\frac{2\pi}{T}\right)^2 = \frac{GM}{r^2}$$

4   Vectors and Scalars

Scalar quantities have a magnitude (size) only, whereas vector quantities have a magnitude and a direction.

Vectors can be represented graphically as arrows. The length of the arrow indicates the magnitude of the vector. The direction of the arrow indicates the direction of the vector.

Quantity	Vector or Scalar?
Distance	Scalar
Time	Scalar
Displacement	Vector
Velocity	Vector
Acceleration	Vector
Speed	Scalar
Force	Vector
Gravitational potential energy	Scalar
Kinetic energy	Scalar
Momentum	Vector

When two vector quantities are added, the two arrows that represent the quantities are joined tip-to-tail:

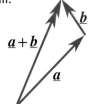

Subtracting a vector is the same as adding a vector pointing in the opposite direction.

If two vectors are in opposite directions, they add to give a vector with magnitude equal to the difference of the original vectors' magnitudes.

$$2\,\text{N} \qquad\qquad 4\,\text{N}$$
$$4\,\text{N} - 2\,\text{N} = 2\,\text{N}$$

If two vectors are at right angles, the sum of their magnitudes can be calculated using Pythagoras' theorem.

4.1   In each example, state the size and direction of the unbalanced force acting on the object. (Forces to the right or upward are assigned to be positive, whilst forces to the left or downward are assigned to be negative.)

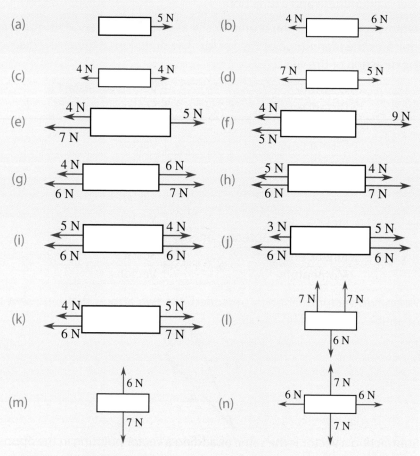

4.2   What is the resultant force on a racing car with 24.5 kN of driving force and 15.2 kN of opposing frictional forces (i.e. drag)?

4.3   What is the resultant magnitude of the displacement if a person walks north 5.00 km and east 4.00 km? i.e. How far away are they from the starting point?

4.4   Why is gravitational field strength a vector quantity?

4.5   Weight is a force. Force is a vector and so it has a direction. In what direction does your own weight point in these situations?

(a) not moving

(b) walking sideways

(c) walking in a circle

4.6   A stunt man drives a car out of the back of a moving lorry. For the stunt to work, the car must be moving at a velocity of -5.00 m/s the instant it has left the lorry. The lorry is travelling at a velocity of 25.0 m/s. What speed must the speedometer on the car reach before the car leaves the lorry?

4.7   Using a scale diagram, calculate the resultant force acting on a sailing boat when an easterly wind provides 2.50 kN of force, the tide provides 1.20 kN of force from the direction $30.0°$ more northerly than the wind.

4.8   A hiker walks 10.0 km east, 5.00 km south and 2.00 km west. Using a scale diagram, calculate his bearing from his start point. *[Note: bearings are given as angles where due north is $0°$ and the angle increases clockwise such that due east is $90.0°$.]*

4.9   A kite is in equilibrium, so the total sum of the forces is equal to zero. On a vector diagram, the arrows representing the forces would form a closed loop. Three forces act on the kite; the force from the wind, the weight of the kite and the tension in the string. The wind produces a horizontal force of 70.0 N and an upward force of 50.0 N and the kite weighs 25.0 N. Use a scale diagram to find:

(a) the tension in the string;

(b) the acute angle the string makes to the horizontal.

5   Variables and Constants

Measurable quantities are either variables or constants. A variable is a quantity whose value can change. A constant is an unchanging quantity.

Commonly used constants include:

charge of the electron	-1.60×10^{-19} C
speed of light in a vacuum	3.00×10^8 m/s

Some quantities *can* have different values (so they are variables), but within a particular experiment we do not expect their value to change. With these quantities, every effort should be taken to make sure their value remains as constant as possible. These are called control variables. Sometimes, deducing a value of a control variable and comparing this to an expected value is a useful way of testing the validity of the experiment. Common control variables include:

gravitational field strength at the surface of the Earth	9.81 N/kg, but taken as 10 N/kg at GCSE level
specific heat capacity of water	4 200 J/(kg °C)
speed of sound in air	330 m/s
refractive index of glass	1.50

In any experiment, the value of one quantity must be systematically changed in order to measure its effect on another quantity. The quantity that the experimenter chooses to change is called the independent variable.

The quantity whose value changes in response to the change of independent variable value is called the dependent variable.

Often, the independent variable and dependent variable values will be plotted on a graph so that the relationship between the two can be deduced and predictions can be made and tested.

5.1   Scientists wish to know the acceleration of a car as it rolls down a sloping ramp. They set the ramp at a certain angle and then release the car from different positions up the ramp, timing how long it takes to reach the bottom. There are several quantities that can be changed in this experiment. For each of the following, state whether it is a control variable, independent variable or dependent variable.

Variable	Variable type
Length of the ramp	(a)
Distance the car rolls	(b)
Duration of the car's motion	(c)
Mass of the car	(d)
Angle of the ramp	(e)
Surface material of the ramp	(f)

5.2   A sportsman wants to know the bouncing efficiency of a table tennis ball. He drops the ball from various heights and measures the maximum height the ball reaches after the first bounce. For each of the quantities listed in the table, state whether it is a control variable, an independent variable or a dependent variable.

Variable	Variable type
Size of ball	(a)
Material of ball	(b)
Height of ball before being dropped	(c)
Maximum height of ball after one bounce	(d)
Mass of ball	(e)
Material of surface onto which ball is dropped	(f)

6 Straight Line Graphs

To be able to correctly predict the effect of changing one variable on the value of another, physicists write equations. Part of the process of writing an equation requires the physicist to draw a graph, which reveals how one variable relates to another. When drawing graphs, it is common practice to plot the independent variable on the x-axis (the horizontal axis), and the dependent variable on the y-axis (the vertical axis). Occasionally, it is more sensible to plot the variables on the axes the other way around. The equation for a straight line graph is:

$$y = mx + c$$

where y is the variable plotted on the y-axis, x is the variable plotted on the x-axis, m is the gradient of the straight line and c is the y-intercept.

At GCSE level, the relationship between two chosen variables is often linear, which means a graph of one variable versus another produces a straight line graph and the above equation works. Most equations at GCSE level can be written in the form $y = mx + c$.

Example – If a student records every second how far something has travelled at constant speed, they can plot a graph distance on the y-axis and time on the x-axis. The gradient will be the speed.

6.1   A student wishes to measure the resistance, R, of a fixed resistor by varying the potential difference, V, across it and measuring the current, I, that flows through it. These quantities are related by $V = IR$. You might find it useful to re-write this relation as $I = (1/R) \times V$ The student plots V on the x axis.

(a) What variable should be plotted on the y-axis?

(b) How can the resistance of the fixed resistor be determined from the graph?

The gradient of a straight line can be determined by considering two points on it:

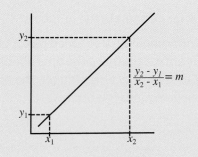

$$m = \frac{y_2 - y_1}{x_2 - x_1} = \frac{(21-7)\ m}{(15-5)\ s} = \frac{14\ m}{10\ s} = 1.4\ m/s$$

6.2   For the following graph, calculate the gradient of the straight line sections labelled a, b, c, d, e, f and g.

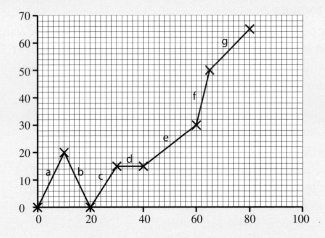

6.3   Write the equation of a line which has a gradient of 2 if $y = 5$ when $x = 0$.

6.4   Write the equation of a line with gradient of 5, if $y = 7$ when $x = 1$.

6.5   Write the equation of a line with gradient of -8, if $y = 0$ when $x = 5$.

7 Proportionality

Physicists measure things, and then look for patterns in the numbers.

The most important pattern is called proportionality (also called direct proportion). If distance is proportional to time, it means that if the time doubles, the distance will double too. If the distance gets 10 times bigger, the time will get 10 times bigger as well. Mathematically, this is written as $s \propto t$.

> **Example 1** – A particular resistor passes a 25 mA current when the voltage across it is 5.5 V. If voltage is proportional to current, what will the voltage be when the current is 60 mA?
>
> The new current is (60 mA / 25 mA) = 2.4 times larger than the old one. The new voltage will be 2.4 times larger than the old one: 5.5 V × 2.4 = 13.2 V.

Using a formula

If s is proportional to t then s/t will always have the same value. If we call this fixed value k, it follows that $k = s/t$, and that $s = kt$. We can use this information to answer questions. The formula method is much clearer if there are more than two quantities involved.

> **Example 2** – A spring obeying Hooke's Law (its extension is proportional to the force) stretches by 14 mm when a 7.0 N load is applied. How far will it stretch with a 3.0 N load?
>
> We write force $= k \times$ extension, so extension $=$ force$/k$.
> $k =$ force/extension $= 7.0\,\text{N}/14\,\text{mm} = 0.50\,\text{N/mm}$
> For a 3.0 N load, extension $=$ force$/k = 3.0/0.50 = 6.0$ mm.

> **Example 3** – The energy transferred by an electric circuit in a fixed time is proportional to the voltage and also to the current ($E \propto V \times I$). If the current is 3.2 A, and the voltage is 15 V and the energy transferred is 340 J. What current will be needed if we need to deliver 640 J using 12 V in the same time?
> The equation is $E = kIV$, so $k = E/(IV) = 340\,\text{J}/(3.2\,\text{A} \times 15\,\text{V}) =$

7.08 J/(AV)
Rearranging gives $I = E/(kV) = 640/(7.08 \times 12) = 7.53 = 7.5$ A
(2sf)

7.1 A cyclist can travel 9.0 km in 30 minutes on level ground. Assume that their speed is constant.

(a) How far will they go in 120 minutes?

(b) How far do they go in 20 minutes?

(c) How much time will it take them to cover 27 km?

(d) How much time will it take them to cover 15 km?

7.2 One day, €1.00 is worth £0.83. On that day

(a) How many pounds would be needed to receive €200 when exchanging your money?

(b) How many euros could I get for £150 (to the nearest €)?

(c) A sandwich in a popular tourist city costs €6.50. How much is that in pounds (to the nearest penny)?

(d) A railway ticket costs £23.50. How much is that in euros (to the nearest cent)?

7.3 The UK minimum wage was £3.87 per hour for someone under the age of 18. Assume that your employer paid you this wage.

(a) How much did you earn for 20 hours of work?

(b) You worked 100 minutes a day after school. How much did you earn a day?

(c) How many hours would you have had to work to save £200?

7.4 The number of widgets made in a factory each week is proportional to the number of workers and the number of hours each worker works. When the factory employs 25 staff, each working 35 hours/week, 65 400 widgets were made each week.

(a) How many widgets would be made each week if 40 staff worked for 30 hours per week?

(b) If we need 130 000 widgets made each week, and the staff will work 42 hours/week, how many workers are needed?

7.5   The merchandiser at a warehouse sends stock to stores in proportion to their sales. She has 670 pairs of mauve trousers to dispatch. Her sales figures tell her that 124 pairs of trousers were sold in total last week, with the New Town branch selling 18 of them. How many pairs of trousers should she send to New Town?

7.6   A watch is set to the correct time at noon on 1st January and put in a drawer. When it is checked at noon on 1st February, it reads 11:51:20. What did it read at 6:00am on 24th January?

Inverse Proportionality

The time taken on a journey is inversely proportional to the speed. If you double the speed, the time halves. If you only go at a tenth of the speed, it takes $10\times$ as long. We write this as $t \propto 1/v$, where v is the speed. In this case, $v \times t$ always has the same value.

Example 4 – The number of books printed each day is proportional to the number of printers owned, and inversely proportional to the number of pages in each book.
If 3 000 300-page books can be printed in one day on 8 printers, how many 125-page books can they print on 6 printers in a day?

As books $\propto$ printers, and books $\propto 1/$pages,
then books $= k \times$ printers$/$pages.
$k =$ books $\times$ pages$/$printers $= 3\,000 \times 300/8 = 112\,500.$
books $= k \times$ printers$/$pages $= 112\,500 \times 6/125 = 5\,400.$

7.7   A cyclist's journey to work takes them 32 minutes at 19 km/h. [*Hint: time $\times$ speed $= \frac{32}{60} \times 19 = 608$.*]

(a) How long would it take at 15 km/h?

(b) How fast would they have to go to reduce the time to 25 minutes?

7.8   An interest free loan for a luxury sofa takes 15 months to pay back at £80/month. Monthly charge $\propto$ number of sofas/duration of loan. What would the monthly charge be if I bought 3 sofas and paid for them over one year?

7.9   The current through a resistor is inversely proportional to its resistance. With a 330 Ω resistor, the current is 25 mA. What value of resistance is needed if you wish a 55 mA current to pass?

7.10   The braking force required to stop a car is inversely proportional to the time taken to stop it. If a 5 500 N force can stop the car in 8.0 s, how much force would be needed to stop it in 3.5 s?

$^{15}/_{20}$

Additional Proportionality Questions

7.11   Which two criteria must be met for a line graph to indicate direct proportionality between two quantities?

7.12   For each of the following equations state whether the two stated variables are directly proportional, inversely proportional or neither. [If there are other values in the question, they are kept constant.]

(a) $W = mg$ - W and m

(b) $pV = kT$ - p and V

(c) $p = mv$ - p and v

(d) $F = k\dfrac{Q_1 Q_2}{r^2}$ - F and r

(e) $T(\text{K}) = T(^{\circ}\text{C}) + 273$ - $T(\text{K})$ and $T(^{\circ}\text{C})$

(f) $a = 4\pi^2 r f^2$ - a and f^2

Mechanics

When we study motion, distance is a scalar quantity that is equal to how far an object has moved. It is measured in metres in SI units. Other units include centimetres, inches, yards, miles and lightyears.

Time is central to the study of motion. It is measured in seconds in SI units. Other units include minutes, hours and days.

Speed is a scalar quantity that is equal to how far an object has moved divided by the time taken. It is measured in metres per second in SI units. Other units include miles per hour, parsecs per jubilee and feet per Julian year: Any speed unit using a distance unit divided by a time unit is valid.

The equation for average speed is:

average speed $=$ total distance$/$total time $[v = s/t]$

In the equation, physicists use v for speed and s for distance. These symbols are useful for more advanced mechanics. Always define your symbols.

Typical speeds are: Walking: $1.5\,\text{m/s}$ Running: $3\,\text{m/s}$ Cycling: $6\,\text{m/s}$

8.1   Use: speed = distance / time to calculate the missing values.

Distance (m)	Time (s)	Speed (m/s)
100	10	(a)
990	3.0	(b)
2.0×10^3	5.0	(c)
(d)	10	330
(e)	5.0	3.0×10^8
3 600	(f)	12
1.2×10^9	(g)	3.0×10^8
1.7×10^4	(h)	340

8.2   Work out the missing measurements from the following table, where each row is a separate question.

Average speed	Total distance	Total time
330 m/s	(a)	10.0 s
(b)	6.00 km	20.0 μs
3.00 m/s	45.0 m	(c)
(d)	40.1 km	24.0 hours
29.8 km/s	940 Gm	(e)
0.047 0 km/h	(f)	2 min 33 s
(g)	100 m	8.13 s

8.3   A train has an average speed of 100 kilometres per hour. Explain why the maximum speed could be different.

8.4   How far can you run in 15 seconds at an average speed of 8.0 m/s?

8.5   How long does a car take to travel 2.4 km at an average speed of 30 m/s?

8.6   A good long distance runner has an average speed of 5.5 m/s. How far would the runner go in 30 minutes?

8.7   The London-Glasgow shuttle takes approximately 60 minutes to fly a distance of 650 km. Estimate its average speed in m/s.

8.8   The wandering albatross can fly at speeds of up to 32 m/s (the speed limit on motorways!). One albatross was found to have flown 16 250 km in 10 days. Calculate its average speed in metres per second.

8.9   A cross-channel ferry travels at about 7 m/s. At the same average speed, how long would it take to cross the Atlantic Ocean, a distance of 6 700 km? Answer to the nearest hour.

8.10 How many kilometres is a 'light-year' – the distance travelled through space in a year by light travelling at 300 million metres per second?

Additional Speed, Distance and Time Questions

8.11   At what speed does a bowler bowl a ball if it travels the length of the wicket to the batsman (20 metres) without bouncing in 0.45 s?

8.12   Concorde had a top speed of around 2 180 km/h; (that is, about twice the speed of sound in air, 340 m/s). Calculate its time to fly across the Atlantic Ocean from London to New York at this speed, a distance of 7 600 km.

8.13   A sock on the rim of a washing machine drum whilst it is spinning goes round in a circular path of radius 20 cm at a rate of 15 times per second. Calculate the speed of the sock in metres per second. Remember that the circumference of a circle, $c = 2\pi r$.

8.14   A marathon race is run over a distance of 42 730 metres. A top runner can complete the course in 2 hours 15 minutes. Calculate the average speed of the runner in metres per second.

8.15   Calculate the speed of a point on the Earth's equator as the Earth rotates once each day. The radius of the Earth is 6400 km.

8.16   Calculate the speed of the Earth in its orbit around the Sun if the radius of the orbit is 1.50×10^{11} m.

8.17   A delivery person starts their delivery round at 6:30am. They travel a total distance of 5.00 km. At 7:53am the delivery round ends. What was their average speed?

8.18   A police constable drives down a motorway with an average speed of 110 kilometres per hour. How far does the police constable travel in 15.0 seconds?

8.19   A groom is marrying his partner at 1:00pm. The wedding venue is 6.00 km away from their house and the average journey speed is 40 kilometres per hour. What is the latest time he can leave his house in order to arrive on time?

9 Displacement and Distance ♡

The straight line distance between an object's starting point and its end point - together with the direction - is called its displacement. The length of the path along which the object moves is the distance.

 Displacement is a vector since it has a direction associated with it. Distance is a scalar; see Section 8.

Distance and displacement are both measured in metres (m) in SI units.

If an object moves in a circle, after one complete rotation, the displacement will equal zero and the distance travelled will equal the circumference of the circle.

If an object is displaced in two perpendicular steps, the magnitude of the displacement can be calculated using Pythagoras' theorem and the direction can be calculated using trigonometry.

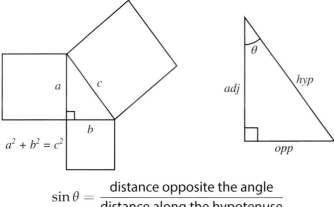

$$\sin \theta = \frac{\text{distance opposite the angle}}{\text{distance along the hypotenuse}}$$

$$\cos \theta = \frac{\text{distance adjacent to the angle}}{\text{distance along the hypotenuse}}$$

$$\tan \theta = \frac{\text{distance opposite the angle}}{\text{distance adjacent to the angle}}$$

9.1 A bus travels 500 m east, 250 m north, 500 m east and 250 m south.
   (a) What distance has the bus travelled?
   (b) What is the final displacement of the bus?

9.2 A climber climbs 50.0 m up a vertical cliff face, before being forced to climb back down 5.00 m so that she can find an alternative route to the top of the cliff. She climbs sideways to the left 10.0 m, then continues to climb 55.0 m to the top.

(a) What distance has the climber travelled?

(b) What is the magnitude (size) of her final displacement measured from her starting point?

9.3 An object is displaced by 60.0 m at a bearing of 60.0°. If the object then moved 30.0 m due north, what is the final magnitude of displacement of the object from its origin?

9.4 A box is dropped from an aeroplane 2 000 m high travelling horizontally at 100 m/s. The box takes 20.2 s to hit the ground. While the box speeds up vertically, it continues at 100 m/s horizontally.

(a) What distance has the box travelled horizontally when it hits the ground?

(b) When the box hits the ground, what is the magnitude of the displacement of the box from the location it was released? i.e. how far is the box from the release point?

(c) What is the angle between the horizontal and the displacement of the box from the location it was released? Give your answer to 3 significant figures.

9.5 A bridge is a quarter of a circle of radius 20.0 m.

(a) What distance does a car travel whilst crossing the bridge?

(b) What is its displacement from the start to the end of the bridge?

9.6 A car is moving at a speed of 10.0 m/s. It is on a roundabout with a diameter of 50 m. After 23.56 s on the roundabout:

(a) What distance has the car travelled?

(b) How many turns of the roundabout has the car made?

(c) What is the magnitude of the car's displacement from where it entered the roundabout?

(d) Other than at the very start, $t = 0$ s, is it possible for the distance the car has traveled to equal its displacement at any point on the roundabout? Explain your answer.

10   Motion Graphs; Displacement–Time (s–t)

A displacement-time graph has displacement on the y-axis (the vertical axis) and time on the x-axis (the horizontal axis). The gradient of the line at any point is the velocity at that instant.
To review gradient calculations, see Straight Line Graphs - P14.

Take particular care of the unit for the gradient. It will be equal to the unit on the y-axis divided by the unit on the x-axis. For example, if displacement is measured in km on the y-axis and time in minutes on the x-axis, the gradient would have units of km per minute.

When displacement is on the y-axis, the direction of the displacement is equal to the direction of the velocity, unless the gradient has a negative value, in which case the direction of the velocity is opposite to the direction of the displacement.

When distance is on the y-axis instead of displacement, the gradient equals speed instead of velocity.

10.1   Describe the motions of the object for which the following displace-
        ment – time graphs have been constructed.

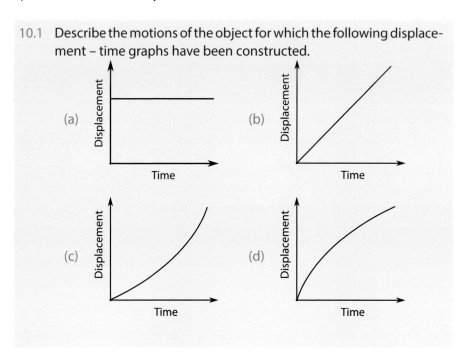

10.2   For the graph below, calculate the velocity for each labeled section
       a – j.

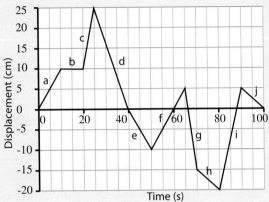

10.3   Considering the graph below:

(a) Between which times is the velocity most negative? Calculate
the velocity between these times.

(b) Between which times is the velocity most positive? Calculate
the velocity between these times.

(c) Between which times is the speed highest? Calculate the speed
between these times.

(d) Between which times is the speed lowest? Calculate the speed
between these times.

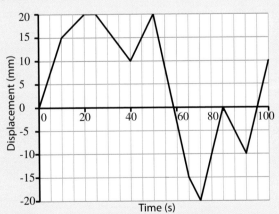

11   Acceleration

Acceleration means that there is a change of velocity – a change of speed or a change of direction of motion.

This could mean

- speeding up
  - when the acceleration is in the same direction as the motion
- slowing down (also called deceleration)
  - here the acceleration is in the opposite direction to the motion
- changing direction (a centripetal acceleration)
  - here the acceleration is at right angles to the motion

We measure acceleration in metres per second squared (m/s^2). An acceleration of 3 m/s^2 means that each second the velocity changes by 3 m/s.

acceleration (m/s^2) = change in velocity (m/s) / time taken (s)

$$a = (v - u)/t$$

When the velocity changes we use u for the velocity at the start, and v for the velocity at the end.

Example 1 – A car is travelling at 3.0 m/s. It accelerates at 2.5 m/s^2. How fast is it going 5.5 s later?

Change in velocity = $a \times t$ = 2.5 m/s^2 × 5.5 s = 13.75 m/s
New velocity = 3.0 + 13.75 = 17 m/s (2sf)

Example 2 – A car at 31 m/s stops in 6.8 s. Calculate the deceleration.

Acceleration = $(v - u)/t$ = (0 m/s − 31 m/s)/(6.8 s) = (−31 m/s)/ (6.8 s) = −4.56 m/s^2 so deceleration = 4.6 m/s^2 (2sf)
Here the velocity change is negative as the final velocity (0 m/s) is lower than the starting velocity (31 m/s), thus is a deceleration.

> **Example 3** – A car starts from rest. It accelerates backwards until it is reversing at 4.0 m/s. This takes 5.0 s. Calculate the acceleration.
>
> Acceleration $= (v - u)/t = (-4.0 \text{ m/s})/(5.0 \text{ s}) = -0.80 \text{ m/s}^2$.
> The change in velocity is negative as the final velocity (-4.0 m/s) is lower than the starting velocity (0 m/s). However, although the acceleration is negative, this is not a deceleration as the car is speeding up (backwards).

11.1 Complete the table with the correct values. Each row represents a separate situation.

Acceleration (m/s²)	Velocity (m/s) after ... s						
	0.0	**1.0**	**2.0**	**3.0**	**4.0**	**5.0**	**6.0**
3.0	0.0	3.0	(a)	9.0	(b)	(c)	18
5.0	0.0			(d)		(e)	(f)
7.0	3.0			(g)	(h)		(i)
-25.0	30.0			(j)	(k)		(l)
(m)		10.5	13.5		(n)		
(o)		45	36	27			(p)

11.2 In Q11.1(d), what would the velocity be 15 s after the start if the acceleration were maintained?

11.3 In Q11.1(o), at what time does the vehicle come to a stop?

11.4 A tennis ball is thrown in the air upwards at 15 m/s. If it is accelerating downwards at 10 m/s², what will its velocity be 2.0 s after it is thrown? (Remember to say how fast it is going and also which way.)

11.5 A rollercoaster speeds up from rest to 100 mph (45 m/s) in 1.2 s.

(a) Calculate the acceleration.

(b) The rollercoaster car then travels vertically upwards, and decelerates at $10\,\text{m/s}^2$. How much time passes before it is stationary (for a moment)?

11.6   A car starts from rest and reaches a speed of 40 m/s in a time of 8.0 seconds. Calculate its average acceleration.

11.7   Complete the table with the correct values. Each row represents a separate situation.

Starting velocity (m/s)	Final velocity (m/s)	Time taken (s)	Acceleration (m/s^2)
0.0	(a)	8.5	3.5
4.5	35	8.5	(b)
26	0.0	(c)	−6.7
(d)	5.0	1.2	−1.5
0.0	(e)	300	31

11.8   A certain make of car can reach 60 mph from rest in a time of 9.0 seconds. Calculate its average acceleration in m/s^2. [Note: 1 mph = 0.45 m/s]

11.9   Calculate the change of speed of a train which accelerates for 9.0 seconds at a rate of $1.2\,\text{m/s}^2$ in a straight line.

11.10 In overtaking a lorry on a straight section of road, a driver increases speed from 50 mph to 70 mph in 5.0 s. [Note: 1 mph = 0.45 m/s.] Calculate the acceleration in:

(a) miles per hour per second and;

(b) metres per second per second.

12   Motion Graphs; Velocity–Time (v–t)

The displacement of an object moving with a constant velocity is equal to the product of the velocity and the amount of time the object is in motion.

To find the displacement when the velocity is changing, a velocity-time graph is needed. Normally, velocity is plotted on the y-axis (the vertical axis) and time is plotted on the x-axis (the horizontal axis).

The area under the line on a velocity-time graph is equal to the displacement of the object.

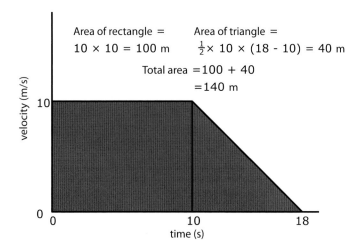

If the shape of the graph can be broken into simple geometric shapes, the total area under the line can be calculated by adding the areas of those shapes.

The area under a speed-time graph is the distance. Speed cannot be negative, and neither can the distance.

The area under a velocity-time graph is the displacement. Velocity can be negative if an object is moving backwards. The displacement can also be negative. An area beneath the x-axis has a negative value. An area above the x-axis has a positive value. Be careful when calculating the total displacement, when summing the displacements remember to include the $+$ and $-$ signs of the displacements.

12.1 Using the following speed–time graph:

  (a) calculate the distance travelled in A;

  (b) calculate the distance travelled in B;

  (c) calculate the distance travelled in C;

  (d) calculate the total distance travelled.

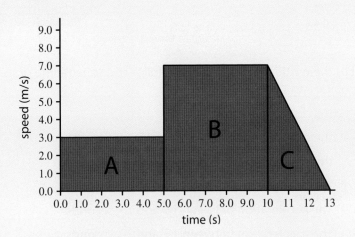

12.2 Using the following graph:

  (a) calculate the displacement in A;

  (b) calculate the displacement in B;

  (c) calculate the displacement in C;

  (d) calculate the displacement in D;

  (e) calculate the total displacement.

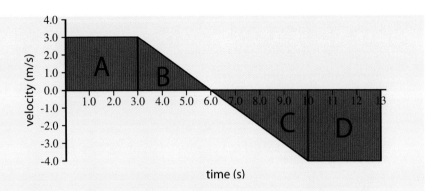

12.3 For the motion described by the following speed–time graph, calculate:

(a) the distance moved in the first 10 s;

(b) the distance moved in the first 15 s;

(c) the total distance moved.

(d) The acceleration between 0 and 10 seconds.

(e) The acceleration between 10 and 15 seconds.

(f) The acceleration between 15 and 20 seconds.

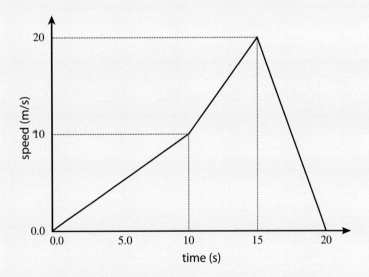

12.4 Calculate the displacement moved and the acceleration for the fol-
       lowing velocity–time graphs.

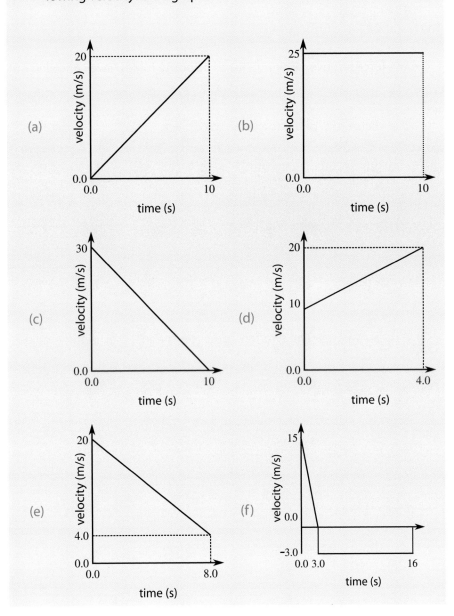

13   Resultant Force and Acceleration

The resultant force on an object is:

- the force left over after equal and opposite forces have cancelled out;

- the one force which would have the same effect as all of the forces;

- the vector sum of the forces on the object.

Example 1 – Calculate the resultant force on this object.

2 N force to left cancels out 2 N of the 6 N of the right force, leaving $6\,\text{N} - 2\,\text{N} = 4\,\text{N}$ to the right left over.

Or you can answer: The two forces are $+6$ N and -2 N. Adding gives 4 N.

Or you can add the vector arrows 'nose to tail' to get a resultant 4 N answer:

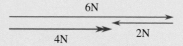

[A double arrow symbol here denotes a resultant vector.]

13.1   For these questions, refer to the diagrams that follow.

(a) Work out the size (strength) and direction of the resultant forces for each object. [Hint for (vi): draw the vectors nose to tail and think 'Pythagoras'.]

(b) Work out the size and direction of the extra force which would need to be added in order to achieve equilibrium (zero resultant force) for each object.

(c) Compare your answers to (a) and (b). What do you notice?

(i)   3.0 N ⟵ (3.0 kg) ⟶ 3.0 N   (ii)   9.0 N ⟵ (3.0 kg) ⟶ 12 N

(iii)   15 N ⟵ (10 kg) ⟶ 20 N   (iv)   3.0 N ⟵ (5.0 kg) ⟶ 3.0 N
                                                                        ↓ 2.0 N

                                                                     (20 kg) ⟶ 3.0 N
(v)   9.0 N ⟵ (20 kg) ⟶ 12 N   (vi)
               ⟵ 11 N   4.0 N ↓

If you need more practice, turn back to Vectors and Scalars - P9 and try to balance the forces in Q4.9.

The acceleration of an object depends on the:

- resultant force acting on the object;

- mass of the object.

Example 2 – Which of these objects will have the greater acceleration?

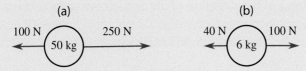

(a)   (b)

100 N ⟵ (50 kg) ⟶ 250 N   40 N ⟵ (6 kg) ⟶ 100 N

(a) has resultant 150 N to the right, acting on 50 kg of mass. This means 150 N/50 kg = 3 N/kg, i.e. 3 N acting on each kilogram.
(b) has resultant 60 N to the right, acting on 6 kg of mass. This means

> 60 N/6 kg $= 10$ N/kg, i.e. 10 N acting on each kilogram.
> Therefore, object (b) will have the greater acceleration.

Formula:

$$\text{acceleration (m/s}^2) = \text{resultant force (N) / mass (kg)} \qquad a = F/m$$

Usually written:

$$\text{resultant force (N)} = \text{mass (kg)} \times \text{acceleration (m/s}^2) \qquad F = ma$$

13.2   Calculate the acceleration of each of the objects in Q13.1.

13.3   Complete the table. Each row represents a different question.

Resultant Force (N)	Mass (kg)	Acceleration (m/s²)
(a)	810	6.7
(b)	430 000	2.6
2 000	65	(c)
(d)	10 g	9.8

13.4   A 100 g mass has weight of 1.00 N.

(a) If this is the only force on the mass, what is its acceleration?

(b) What would be the weight of a 300 g mass in the same gravitational field?

(c) If the weight is the only force on the 300 g mass, what is its acceleration?

A resultant force in the direction of motion speeds an object up.
A resultant force opposite to the direction of motion slows it down.
Zero resultant force means that the object keeps a steady velocity.

13.5   Complete the table. Each row describes a different object which
has two forces acting upon it– one forwards (in the direction of motion), one backwards. Define forces and accelerations acting forwards as positive. Is each object speeding up or slowing down?

Force (N)			Mass (kg)	Acceleration (m/s^2)
Forwards	Backwards	Resultant		
58	16	(a)	5.6	(b)
90	145	(c)	22	(d)
(e)	350	(f)	120	+6.7

13.6 What unbalanced force acts on a 70 kg mass accelerating at 1.6 m/s^2?

13.7 What is the acceleration of a 10 kg mass which has no unbalanced force acting on it?

13.8 A 1 200 kg vehicle is accelerating along a straight road at 3.0 m/s^2. What is the magnitude of the unbalanced force acting on it?

13.9 What force must I apply to a mass of 3.0 kg to accelerate it at 4.0 m/s^2 on a horizontal surface if

(a) there is no friction and;

(b) there is friction of 4.0 N?

13.10 The thrust generated by a rocket engine is equal to the mass of propellant burnt each second multiplied by the exhaust velocity of the gas. The Space Shuttle (with booster rockets and external tank) had a total mass of 2 040 000 kg at launch. In this question we shall assume that the exhaust velocity of the gas was 3 000 m/s.

(a) How much propellant would have to be burnt each second in order for the spacecraft to just lift off?

(b) How much propellant would have to be burnt each second in order for the spacecraft to accelerate upwards from the launch pad at "3g" (i.e. 30 m/s^2)?

$^{34}/_{44}$

Additional Resultant Force and Acceleration – on-line

isaacphysics.org/gameboards#phys_book_gcse_ch_2_13_add

14   Terminal Velocity

A falling object in the air, which is not influenced by wind or other sideways forces, has a maximum of two forces acting on it: weight and air resistance (also known as drag). The weight does not change. The air resistance is zero when the object is stationary but increases as the object speeds up.

The resultant (net) force the falling object experiences is equal to the force of gravity minus the air resistance. [$F = mg - Drag$]

Newton's Second Law states that the acceleration of the object is proportional to the resultant force on the object when its mass is constant.

- The longer the object falls for, the faster it travels (due to its acceleration),

- but then the greater the air resistance, which increases with speed.

- Eventually, however, the air resistance upwards will equal the force of gravity downwards.

- At this point, the resultant force is zero,

- hence, the velocity remains constant. This is called the terminal velocity of the object.

An object with a constant force acting on it in the direction it is travelling and a frictional force (related to the object's velocity) acting in the opposite direction, will reach terminal velocity given enough time. This is true for cars driving along a road or an anchor falling through water towards the seafloor.

14.1   A parachutist of mass 80 kg is falling to the ground at a steady vertical velocity of 10 m/s. What is the value of the total drag force acting on him?

14.2   A bicycle is ridden along a horizontal road with a driving force of 400 N. Its speed is constant at 12 m/s. What is the magnitude of the sum of the frictional forces acting on the bike and its rider?

14.3   Two identical bottles are dropped off a cliff, both falling with their bases downwards. One of them is full of water, while the other contains air at atmospheric pressure. What can you say about the terminal velocities these two bottles will reach?

14.4 A car travelling at constant speed has a driving force of 2.1 kN acting on it. The driver presses the accelerator, and the driving force increases to 2.5 kN.

(a) Immediately after the driver presses the pedal, what is the total resistive force acting on the car?

(b) When the car again reaches a constant speed, what is the total resistive force acting on the car?

14.5 A care package has a mass of 150 kg. It is dropped from a helicopter 2 000 m above the ground.

(a) What is the weight of the care package?

(b) If the object is falling at terminal velocity, what is the value of the air resistance?

(c) The care package has a parachute built in, which opens automatically at 1 000 m. What happens to the air resistance as it opens?

(d) After a short time, the care package plus the open parachute begin to fall with their terminal velocity. How does this value qualitatively compare to the terminal velocity of the care package without the parachute?

(e) How does the air resistance of the care package falling at terminal velocity with the parachute open compare to the air resistance of the care package falling at terminal velocity without the parachute? Explain your answer.

(f) The care package falls off course and lands in a lake. Ignore the effect of buoyancy. The care package starts to sink and reaches terminal velocity once more. How does this value of terminal velocity qualitatively compare to the previous two values? Explain your answer.

(g) In (f), what is the value of the fluid drag through the water when the object sinks at its terminal velocity?

14.6 The online graph shows how the drag force on a 25 000 kg lorry depends on the lorry's speed.

(a) At one moment, the lorry is travelling at 20 m/s and the driving force acting on it is 80 000 N. What is the acceleration of the lorry at this instant?

(b) The lorry maintains a constant driving force of 80 000 N. What is its terminal velocity?

(c) While at the above terminal speed, the driver then halves the driving force to 40,000 N. What is the initial acceleration of the lorry?

(d) After some time the lorry reaches a new terminal velocity. What is this new terminal velocity?

14.7 When an object is travelling through a fluid (liquid or gas), the resistive force will depend on the speed of the object. However, how this force changes with speed depends on whether the flow of the fluid past the object is smooth or turbulent.

(a) If the flow is smooth, the resistive force is proportional to the speed of the object moving through the fluid. If the driving force is multiplied by 5, by what factor does the terminal velocity of the object increase?

(b) If the flow is turbulent, the resistive force is proportional to the square of the speed of the object moving through the fluid. If the driving force is multiplied by 5 , by what factor does the terminal velocity of the object increase?

14.8 When a ship is moving through the water, the resistance to its motion is made up of a few different components. One of these comes from the fact that the ship will push water in such a way to make waves, so it is called the "Wave Making Resistance". The main part of the resistance is known as "Frictional Resistance". For one ship, this provides 80 % of the total resistance force. Give an equation for F_D , the driving force propelling the ship, in terms of the frictional resistance F_f , when the ship is travelling at its terminal velocity.

15   Stopping With and Without Brakes

Formulae:

$$\text{distance travelled} = \text{average speed} \times \text{time} \qquad s = vt$$
$$\text{resultant force} = \text{mass} \times \text{acceleration} \qquad F = ma$$
$$\text{change in velocity} = \text{acceleration} \times \text{time} \qquad v - u = at$$
$$\text{kinetic energy} = \tfrac{1}{2} \times \text{mass} \times \text{speed}^2 \qquad E = \tfrac{1}{2}mv^2$$
$$\text{work done} = \text{force} \times \text{distance} \qquad W = Fs$$

Data:

To convert miles/hr to m/s, multiply by $1\,609/3\,600 = 0.447$.

With Brakes

The shortest distance taken to stop a car from the moment when the driver first notices a problem is called the stopping distance. This is made of two parts – the distance the car travels while the driver reacts and first applies the brakes (the thinking distance), and the distance the brakes take to stop the car (the braking distance).

The Highway Code estimates that a typical reaction time of a driver is two thirds of a second; and that once applied, brakes will give a car a 6.67 m/s^2 deceleration. This reaction time may seem very long - but it takes into account the fact that during a long drive a driver may not be fully alert, and that the action of moving your foot from the accelerator to the brake pedal and stamping takes longer than pressing a button with your finger.

Example 1 – Calculate the thinking distance at 30 mph.

Conversion: 30 mph = $30 \times 0.447 = 13.4$ m/s.
Thinking distance = reaction time $\times$ speed = 0.667 s $\times$ 13.4 m/s = 8.9 m.

> **Example 2** – Calculate the braking distance from 30 mph.
>
> Conversion: 30 mph $= 30 \times 0.447 = 13.4$ m/s.
> Velocity drop $=$ deceleration $\times$ braking time, so
> braking time $=$ velocity reduction / deceleration $= 13.4/6.67 = 2.0$ s.
> Average speed on decelerating from 13.4 m/s to 0 m/s is $(13.4 + 0) / 2$
> $= 6.7$ m/s.
> Braking distance $=$ braking time $\times$ average speed $= 2.0$ s $\times$ 6.7 m/s
> $= 13.4$ m.

15.1 Using the data in the examples, calculate the overall stopping distance from 30 mph (according to the Highway Code).

15.2 Using the Highway code values for reaction time and deceleration:

(a) Calculate the thinking distance at 60 mph.

(b) Calculate the braking distance at 60 mph.

(c) Use Example 1 and your answer to (a) to complete the sentence: when you double your speed, the thinking distance _____.

(d) Use Example 2 and your answer to (b) to complete the sentence: when you double your speed, the braking distance _____.

In your answer to (c), going at twice the speed, you cover twice the distance during your reaction time, as the reaction time doesn't change.
In your answer to (d), going at twice the speed, it takes you twice the time to stop. However, you are going twice as fast, leading to an overall multiplication by 4.

15.3 A car is traveling at 70 mph.

(a) Calculate the thinking distance.

(b) Calculate the braking distance.

(c) Another car is travelling at 35 mph, what is its overall stopping distance?

15.4 How much longer is the stopping distance at 35 mph compared to

30 mph? Measure out this distance. Discuss why the 30 mph speed limit is so important in built-up areas.

15.5 Here is an alternative way of calculating the braking distance from 60 mph (26.8 m/s). We are in a 700 kg car. Calculate:

(a) The kinetic energy of the car.

(b) The force needed for a 6.67 m/s^2 deceleration.

(c) The braking distance. [energy transfer $=$ force $\times$ distance]

15.6 The "two second rule": A driver waits until the vehicle in front passes a road sign and then sees if her car passes this road sign within two seconds. If it has, she is driving too close. Assume that a driver is keeping the two second rule exactly.

(a) How far will the driver be behind the vehicle in front at 30 mph?

(b) How far will the driver be behind the vehicle in front at 60 mph?

(c) By referring back to your answer to Q15.1, does a driver following the 2 second rule at 30 mph always keep at least one stopping distance behind the vehicle in front?

(d) By referring back to your answer to Q15.2, does a driver following the 2 second rule at 60 mph always keep at least one stopping distance behind the vehicle in front?

$^{16}/_{20}$

In wet conditions, drivers should allow *at least* twice these distances.

Without Brakes

15.7 A car runs into a wall and stops in 0.30 s. It was going at 20 m/s.

(a) Calculate the deceleration.

(b) A person fixed to the car by a seatbelt has the same deceleration. They have a mass of 70 kg. Calculate the force on the person.

(c) Repeat the calculation for the force if the car took 0.90 s to stop.

16   Moments, Turning and Balancing

The turning or twisting effect of a force is called its moment. The moment of a force depends on:

- the size of the force;

- how far the force is from the pivot or axle (the distance is measured from the pivot to the line of action of the force, at right angles to the force);

- the direction of the force. Moments can be anticlockwise (AC) or clockwise (C).

moment (Nm) = force (N) × perpendicular distance to axle (m)

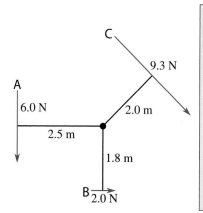

Example 1
Moments of the forces in the diagram are:
A: 6 N × 2.5 m = 15 Nm AC
B: 2 N × 1.8 m = 3.6 Nm AC
C: 9.3 N × 2 m = 18.6 Nm C
In this case, the two AC moments added together equal the C moment. Because they are equal and in opposite directions, the system will not turn.

Principle of Moments:
An object will balance and not start turning if the total of the anticlockwise (AC) moments equals the total of the clockwise (C) moments.

16.1 Calculate the missing values in the table. Each row is a different question. Give your answers in the units requested.

	Force	Distance from axle	Moment
(a)	4.2 N	5.4 m	(Nm)
(b)	68 N	0.15 m	(Nm)
(c)	47 N	34 cm	(Ncm)
(d)	(N)	3.2 m	18 Nm
(e)	0.034 N	(cm)	0.68 Nm

16.2 What force, acting at a distance of 25 cm from the axis of rotation of a solid body, would make a moment of 10 Nm?

16.3 What force, acting at a distance of 2.5 m from the axis of rotation of a solid body, would make a moment of 25 Nm?

16.4 Two forces act on a rigid body free to rotate about a perpendicular axis through point O. The sizes of the forces are 10 N and 20 N.

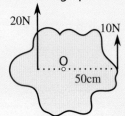

The perpendicular distance of the line of action of the 10 N force from O is 50 cm. What is the perpendicular distance of the 20 N force from O if the body does not rotate?

16.5 Two forces act on a rigid body free to rotate about an axis perpendicular to the point O. One of the forces is 20 N and acts at a perpendicular distance of 60 cm from the axis.

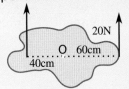

If the other force acts at a perpendicular distance of 40 cm from the axis and the body is in equilibrium, what is the size of the force?

16.6 For the situations below, work out the moment of each force and state whether the object will begin to turn anticlockwise (AC), clockwise (C), or whether it is balanced (B). Draw a diagram of each situation to help you decide whether each moment is AC or C.

(a) A 3.2 N force to the left is 10 cm above the axle, while a 6.4 N force to the left is 5.0 cm below the axle.

(b) A downwards 65 N force is 35 cm to the left of an axle, and a 150 N force to the right is 20 cm above the axle.

(c) An upwards 650 N force is 2.3 m to the right of an axle, while 150 N to the left is 3.6 m below the axle.

(d) A force of 10 N to the right is 2.3 m to the right of the axle, and a force of 0.20 N upwards is 0.34 m to the right of the axle.

(e) An upwards 30 N force is 35 cm to the left of an axle, while an upwards force of 15 N is 70 cm to the right of the axle.

(f) Two upwards forces act: one of 34 N which is 3.5 m to the left of the axle, the other is 25 N and is 2.6 m to the right of the axle. There is also a 10 N force to the left which is 0.80 m below the axle.

Example 2 – What force, F, is needed to make the rod balance?

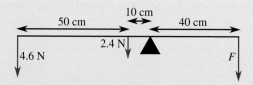

Moment of a 4.6 N force is 4.6 N × 60 cm = 276 Ncm AC
Moment of 2.4 N force is 2.4 N × 10 cm = 24 Ncm AC
Total AC moment is 300 Ncm. To balance, the moment of F must be 300 Ncm clockwise, so
$F = 300$ Ncm$/40$ cm $= 7.5$ N
In this question, the 2.4 N force is the weight of the rod. Weights are always drawn downwards from the centre of gravity - which is always in the centre of symmetric, uniform objects like rods.

16.7 Calculate the weight of the block stated in each situation below where the uniform lever arm is balanced about the fulcrum 'F'.

(a) If A weighs 5.0 N, what is the weight of B?

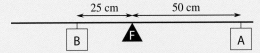

(b) If A weighs 10 N, what is the weight of B?

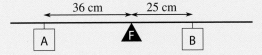

(c) If A weighs 10 N, what is the weight of B?

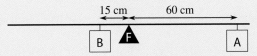

(d) If A weighs 10 N and B weighs 20 N, what is the weight of C?

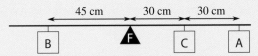

(e) If A weighs 2.0 N and B weighs 4.0 N, what is the weight of C?

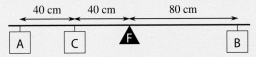

16.8 A 0.50 N weight is stuck to the 20 cm mark of a uniform metre stick, which weighed 0.50 N before the weight was added. You can balance the metre stick horizontally on your finger, if you put your finger in the right place. How far from the 0.0 cm end do you need to put your finger in order to get it to balance?

17   Pressure, Hydraulic Systems, Density and Depth

$$\text{pressure} = \text{force (N)}/\text{area (m}^2) \qquad p = F/A$$

The unit of pressure is the pascal (Pa). $1\,\text{Pa} = 1\,\text{N/m}^2$

Example 1 – What is the pressure on a wall when a drawing pin, with a point of cross sectional area of $2.0\,\text{mm}^2$, is pushed in with a force of 8.0 N?

Pressure $= \text{force}/\text{area} = 8.0\,\text{N}/2.0\,\text{mm}^2 = 4.0\,\text{N/mm}^2$.

Notice that $1\,\text{mm}^2 = 1\,\text{mm} \times 1\,\text{mm} = 10^{-3}\,\text{m} \times 10^{-3}\,\text{m} = 10^{-6}\,\text{m}^2$.

Pressure $= \text{force (N)}/\text{area (m}^2) = 8.0\,\text{N}/2 \times 10^{-6}\,\text{m}^2 = 4 \times 10^6\,\text{Pa}$.

17.1   Calculate the value of the missing quantities in the table.

Pressure (Pa)	Force (N)	Area (m²)
(a)	10	2.0
(b)	1.0	4.0×10^{-5}
(c)	10^6	2.5×10^{-5}
10^5	400	(d)
5.0×10^6	2 000	(e)
4.0×10^{-3}	8.0×10^{-3}	(f)
2.5×10^5	(g)	0.020
10^7	(h)	10^{-3}

17.2   Assume the flat end of the drawing pin in Example 1 has an area of $1.2\,\text{cm}^2$. Calculate the pressure on the person's finger who is pushing in the nail with a force of 8.0 N in

(a) N/cm²;   (b) Pa.

17.3   My weight is 670 N, and each of my shoes has a sole area of $200\,\text{cm}^2$.

(a) What will be the pressure when I stand on the ground?

(b) A plank 15 cm wide and 1.5 m long is laid across a muddy path. What will be the pressure on the mud when I stand on the plank?

(c) Compare your answers for (a) and (b). How and why do you think the plank affects whether I sink into the ground?

Pressure in fluids

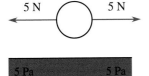

A solid object will not accelerate if the force pulling from each side is equal.

A section of a fluid (liquid or gas) will not accelerate if the pressure on both sides is equal.

In a hydraulic system, two pistons of different area push on the same fluid, exerting pressures on it. The fluid is in equilibrium if the pressures are equal.

The pressure on the left = 4 N/2.5 cm^2 = 1.6 N/cm^2.
In equilibrium, the pressure on the right will also be 1.6 N/cm^2, so the force on the right must be 1.6 N/cm^2 × 40 cm^2 = 64 N.

17.4 Complete the table of hydraulic systems in equilibrium.

Left Piston		Pressure	Right Piston	
Force	Area		Force	Area
3.0 N	0.60 cm^2	(a)	(b)	3.6 cm^2
65 N	15 cm^2	(c)	(d)	0.50 cm^2
45 N	4.5 cm^2	(e)	(f)	25 cm^2
10 N	1.0 cm^2	(g)	(h)	1.0 m^2
35 N	25 cm^2	(i)	7 000 N	(j)

17.5 At a garage, a car (8.0 kN weight) is going to be lifted on four hydraulic jacks, each with a cross sectional area of 25 cm^2. Fluid is forced into the jacks by a compressor. If you want to support the car on the jacks, what is the pressure in the fluid?

Density   density = mass/volume $\rho = m/V$

Density gives the mass of a material per cubic metre (or cubic centimetre). 1.00 kg of water has a volume of 0.00100 m^3. density = mass / volume = 1.00 kg / 0.00100 m^3 = 1000 kg/m^3.

17.6   Complete the table using the formula for density. Give answers in the units requested, shown in brackets in the column headings.

Substance	Water (g/cm^3)	Gold (g/cm^3)	Iron (g)	Ice (kg)	Nitrogen gas (m^3)
Density	(a)	(b)	7.87 g/cm^3	920 kg/m^3	1.13 kg/m^3
Mass	250 g	5.4 g	(c)	(d)	2000 g
Volume	250 cm^3	0.28 cm^3	300 cm^3	434 cm^3	(e)

Remember that 1 cm^3 = $(0.01 \text{ m})^3$ = $(10^{-2} \text{ m})^3$ = 10^{-6} m^3.

Pressure at Depth

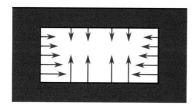

As you go deeper in a fluid, the pressure rises because of the increased weight of fluid above you. However, any surface in the fluid has a force on it regardless of its angle. A box held under water has forces on it from all sides, all pushing inwards at right angles to each surface.

The formula for the extra pressure at a depth is:

  pressure = density × gravitational field strength × depth $p = \rho gh$

To calculate the total pressure at that depth, the pressure at the surface (e.g. atmospheric pressure) must be added.

Example 2 – Calculate the total pressure at a depth of 8.0 m in oil of density 850 kg/m^3 if atmospheric pressure is 101 kPa.

Extra pressure = ρgh = 850 kg/m^3 × 10 N/kg × 0.8 m = 68 000 Pa = 68 kPa

Total pressure = pressure at surface + 68 kPa = 101 kPa + 68 kPa = 169 kPa

---

For these questions assume water has a density of $1\,000\,\text{kg/m}^3$, the gravitational field strength is 10 N/kg and atmospheric pressure is 101 kPa.

17.7 A beaker has a cross sectional area of $0.080\,\text{m}^2$ and is filled to a depth of $0.12\,\text{m}$.

   (a) Calculate the volume of water in m^3.

   (b) Calculate the mass of water in kg.

   (c) Calculate the weight of water in N.

   (d) Calculate the pressure of the water on the base in Pa.

   (e) What would your answers to the previous parts be if the beaker had a cross sectional area of $0.80\,\text{m}^2$, but the same depth of water?

17.8 A watch states that it is 'water resistant to 30 m'.

   (a) What extra pressure can it withstand before leaking?

   (b) What is the extra pressure on the watch at a depth of 10 m?

17.9 The deepest part of the Pacific Ocean, the Mariana Trench, has a depth of $10\,994\,\text{m}$. The density of sea water is $1\,030\,\text{kg/m}^3$.

   (a) What is the total pressure at that depth?

   (b) What would be the inwards force on a 10 cm by 10 cm window in a submarine at the pressure calculated in (a)?

17.10 Mercury has a density of $13\,600\,\text{kg/m}^3$.

   (a) What depth of mercury gives an extra pressure equal to atmospheric pressure?

   (b) At what depth in mercury would the pressure be the same as at a depth of 68 cm in water?

$^{27}/_{36}$

Additional Pressure, Density and Depth Questions

17.11 An ocean temperature probe is lowered from a survey ship into the water. The maximum pressure that the probe is designed to withstand is 100 MPa. What is the greatest depth to which the probe could be safely lowered? The density of sea water is $1\,030\,\text{kg/m}^3$.

18   Moving in a Circle ♡

I swing a bung around in a horizontal circle above my head using a string. From above, it looks like this, and we can ignore the effect of gravity.

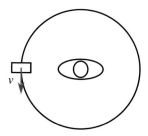

There is a force from the string acting on the bung (not shown in the diagram) acting towards the centre of the circle. Redraw the diagram above in your book to include this force.

The bung is neither speeding up nor slowing down, yet there is an unbalanced force acting on it. This force must therefore be changing the direction of the bung's motion.

- The bung's direction is changing so that it can go round in a circle.
- This means that its velocity is changing, so it must be accelerating.
- This acceleration requires a force.

Any force which causes something to go round in a circle can be labelled as a centripetal force. Three factors which affect the centripetal force are:

- mass of the object;

- its speed;

- the radius of the circle.

Formulae:

$$\text{centripetal acceleration} = \frac{\text{speed}^2}{\text{radius}} \qquad a = \frac{v^2}{r}$$

$$\text{centripetal force} = \text{mass} \times \text{acceleration} \qquad F = \frac{mv^2}{r}$$

The orbit of the Earth around the Sun is approximately circular. The force holding the Earth in this motion is the gravitational force, which acts as a centripetal force in this case.

Draw another arrow on the diagram in your book to show where the bung would go next if the string were cut whilst the bung is at the position shown.

18.1 A 0.50 kg object travels in a 1.4 m radius circle at a speed of 2.0 m/s.

(a) Calculate the force needed to keep it in this motion.

(b) Calculate the force needed for a circle of twice the radius with the same speed. What do you notice?

(c) Calculate the force needed for a speed twice as great for the original circle. What do you notice?

18.2 (a) Rearrange your 'centripetal acceleration' formula to make speed the subject.

(b) How fast are you travelling around a 4.0 m radius circle if the centripetal acceleration is 10 m/s^2?

18.3 What force is needed to enable a 1 525 kg car travelling at 20 mph (8.9 m/s) to go round a roundabout with an 8.0 m radius?

18.4 On a fairground ride, the passengers stand against a railing 5.3 m from the centre of a large wheel. If the wheel rotates once every 2.3 s, what is the acceleration of the riders? [Hint: calculate the speed of the riders first.]

18.5 A 200 kg satellite is going to orbit the Earth at a distance of 6 600 km from the Earth's centre. At this height (200 km from the surface), the gravitational field strength of the Earth is roughly the same as on the surface of the Earth (10 N/kg).

(a) Calculate the weight of the satellite in newtons.

(b) Calculate the speed at which the satellite will need to travel if the centripetal force is to be equal to its weight.

(c) Repeat (a) and (b) for a 120 kg satellite orbiting at the same height.

(d) Repeat (a) and (b) for a 100 kg satellite orbiting at a height of 1 000 km above the Earth's surface. At this height the gravitational field strength is 7.3 N/kg.

(e) Calculate the circumference of both the original orbit and the orbit in (d). For these circumferences, work out the time each satellite would take to orbit once.

18.6 A road over a humped-back bridge can be represented as a sector of a circle of radius 30 m. How fast could you travel over the top of the bridge before your wheels lifted off the ground there?

18.7 The formula for the gravitational field strength g (in N/kg) at distance r (in metres) from the Earth's centre is:

$$g = \frac{k}{r^2}$$

where $k = 4.0 \times 10^{14}$ N m²/kg

(a) Calculate the orbital speed of a satellite orbiting in a circle of radius 4.0×10^7 m around the Earth.

(b) Calculate the time taken for this satellite to orbit once.

(c) Derive a formula for the time for one orbit of a satellite around the Earth in terms of the radius of its orbit. Use k rather than a numeric value in your formula.

(d) At what radius of orbit will a satellite orbit once every 24 hours? Such a satellite is called a geostationary satellite.

(e) The satellites which provide us with satellite TV are geostationary. Why are the TV companies willing to spend the extra money to put a satellite so far away from the Earth, when a nearer one would be much cheaper?

19   Introducing Momentum and Impulse

Momentum measures how much 'motion' an object has, taking into account its mass and velocity.

$$\text{momentum} = \text{mass (kg)} \times \text{velocity (m/s)} \qquad p = mv$$

The unit of momentum is kilogram metres per second (kg m/s)

The sign of the momentum (plus or minus) tells you the direction.
In these one dimensional problems, positive momentum means 'travelling East' and negative momentum means 'travelling West'.

Momentum is a vector – it has a direction.

19.1   Complete the table. Each row represents a separate situation.

Mass	Velocity (m/s)	Momentum (kg m/s)
2.0 kg	+4.5	(a)
1.6 kg	−3.4	(b)
50 g	2.5 East	(c)
60.3 kg	31 West	(d)
120 kg	Stationary	(e)
360 kg	(f)	1 200 East
2.0 g	(g)	2.0 West

Example 1 – A 3.0 kg motion trolley is moving at 2.0 m/s East. A force of 4.2 N acts on it for 6.0 s.
Acceleration = force/mass = 4.2 N/3.0 kg = 1.4 m/s^2
Velocity change = acceleration × time = 1.4 m/s^2 × 6.0 s = 8.4 m/s
New velocity = 2.0 m/s + 8.4 m/s = 10.4 East m/s
Original momentum = mass × velocity = 3.0 kg × 2.0 m/s = +6.0 kg m/s
New momentum = 3.0 kg × 10.4 m/s = +31.2 kg m/s
Change in momentum = 31.2 kg m/s − 6.0 kg m/s = +25.2 kg m/s
Notice that force × time = 4.2 N × 6.0 s = +25.2 Ns

The last line of the example suggests:

$$\text{change in momentum (kg m/s)} = \text{force (N)} \times \text{time (s)}$$

$$p_{\text{after}} - p_{\text{before}} = Ft$$

19.2 A 5.0 kg trolley is initially moving at 3.5 m/s West. A 12.4 N force
      (East) acts on it for 3.5 s. Take 'travelling East' as being positive, and
      'travelling West' as being negative.
      (a) Calculate the acceleration
      (b) Calculate the velocity change
      (c) Calculate the original momentum
      (d) Calculate the new momentum
      (e) Calculate the change in momentum
      (f) Is the change in momentum equal to the product of the force
           and time?

19.3 Complete the table. Each row represents a separate situation. You
      should fill the different columns in the easiest order (which may not
      be left to right). The first row has been worked as an example.

m (kg)	v (m/s)		p (kg m/s)			F (N)	t (s)
	Before	After	Before	Change	After		
1.0	0.0	180	0.0	180	180	3.0	60
2.5	0.0	(a)	(b)	(c)	(d)	4.2	12
25	0.0	(e)	(f)	(g)	(h)	16.9	300
700	10	31	(i)	(j)	(k)	(l)	12
1 800	13	0.0	(m)	(n)	(o)	−12 000	(p)
15 g	0.0	250	(q)	(r)	(s)	(t)	2.0 ms

Example 2 - First row of table above
Momentum before $= mu = 1.0$ kg $\times 0.0 = 0.0$ kg m/s
Momentum change $= Ft = 3.0$ N $\times 60$ s $= 180$ kg m/s
Momentum afterwards $= 0.0 + 180 = 180$ kg m/s
Velocity afterwards $=$ momentum/mass $= 180/1.0 = 180$ m/s

Impulse

We define   impulse (Ns) = force (N) × time (s)

so   impulse (Ns) = change in momentum (kg m/s)

So, a moving object with 400 kg m/s of momentum would need a 400 N force to stop it in one second.

Newton's 2nd Law: resultant force = rate of change of momentum.

19.4 What magnitude of force is needed to accelerate a 300 000 kg wide-body jet from 0.0 m/s to take off speed of 90 m/s in 50 s?

19.5 What will the momentum of a 200 kg rocket be after a 10 kN force has pushed it for four minutes?

19.6 At what speed is a 20 gram air rifle pellet moving if it has a momentum of 1.6 kg m/s?

19.7 A girl on a 10 kg bicycle is riding it at a speed of 6.0 m/s. If the momentum of the girl and bicycle is 360 kg m/s, what is the mass of the girl?

19.8 Your mass is 60.6 kg and you are about to land on your feet after a jump, falling at 0.85 m/s.
Calculate the force on each leg if:

(a) you bend your knees and stop in 0.75 s;

(b) you keep your knees locked and stop in 0.082 s.

19.9 An 800 kg car is travelling at 70 mph and overtaking a 15 000 kg truck travelling at 55 mph. Calculate the ratio of the momentum of the truck to the momentum of the car (p_{truck}/p_{car}).

19.10 Calculate the momentum of a 20 000 tonne ship moving through the water at a speed of 12 m/s. [Note: 1 tonne = 1000 kg]

$^{28}/_{37}$

Additional Introducing Momentum and Impulse Questions

19.11 A car is travelling at 15 m/s. It has 18 000 kg m/s of momentum. What is the car's mass?

19.12 Two cars are travelling in the same direction. One has a mass of 1 000 kg and is moving at 10 m/s, the other's mass is 1 200 kg and it is moving at 15 m/s. What is the total momentum of the cars?

20   Momentum Conservation

momentum (kg m/s) = mass (kg) $\times$ velocity (m/s) $p = mv$

change in momentum (kg m/s) = force (N) $\times$ time (s) $p_{after} - p_{before} = Ft$

The sign of the momentum (plus or minus) tells you the direction.
In these one dimensional problems, positive momentum means 'travelling East' and negative momentum means 'travelling West'.

Example 1

4.5 m/s

1.5 kg ⊣₩₩⊢ 3.0 kg

Two motion trolleys are moving East. The spring expands, and pushes the trolleys apart.   It pushes the 3.0 kg trolley forwards with a 2.5 N force for 1.2 s. What is the change in momentum change of each trolley?

Momentum change of 3.0 kg trolley = Ft = 2.5 N $\times$ 1.2 s = 3.0 kg m/s

By Newton's 3rd Law, force on 1.5 kg trolley must be 2.5 N West (-2.5 N).

Momentum change of 1.5 kg trolley = Ft = -2.5 N $\times$ 1.2 s = -3.0 kg m/s.

The momentum gained by the 3.0 kg trolley is equal to the momentum lost by the 1.5 kg trolley. So the total momentum stays the same.

So, when forces act between objects, their total momentum is conserved.

Example 2 – Calculate the new velocity of the 1.5 kg trolley.
Momentum = old momentum + change = $1.5 \times 4.5 - 3 = 3.75$ kg m/s
New velocity = momentum / mass = $3.75 / 1.5 = 2.5$ m/s (2.5 m/s East)

Example 3 – A 2.5 kg mass travelling at 2.5 m/s collides with and sticks to a 7.5 kg mass which is stationary. Calculate the velocity afterwards.
Total initial momentum: 2.5 kg $\times$ 2.5 m/s + 7.5 kg $\times$ 0 m/s = 6.25 kg m/s
Total final momentum must be the same = 6.25 kg m/s
Final velocity = momentum / mass = 6.25 / 10.0 = 0.63 m/s (2 sf)

20.1 Calculate the momentum of:

(a) a 3.0 kg trolley moving at 2.0 m/s to the East;

(b) a 700 kg car moving at 6.0 m/s to the West;

(c) a 50 g mass moving at 50 cm/s to the East;

(d) a 10 000 kg bus moving Eastwards at 3.0 m/s.

(e) What is the total momentum of the car in (b) and the bus in (d)?

(f) If the car and the bus were to collide and stick together in a crumpled mess, what would the total mass of the wreckage be just after the impact?

(g) What would the total momentum of the crumpled mass be?

(h) Calculate the initial velocity of the wreckage.

20.2 Two 2.0 kg trolleys collide and stick together on a smooth, horizontal surface. One trolley is at rest before the collision.

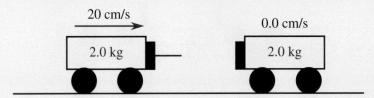

Calculate the combined velocity of the trolleys after the collision.

20.3 Two trolleys are moving in the same direction along a smooth surface. One is moving faster and catches up on the other. The trolleys collide and stick together.

Calculate the combined velocity of the trolleys after the collision.

20.4 Two trolleys are at rest and in contact on a smooth, level surface. A coiled spring in one trolley is released so that they 'explode' apart. The lighter trolley moves off at 50 cm/s.

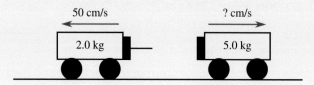

(a) Calculate the speed of the other trolley.

(b) Calculate the minimum energy which was stored in the coiled spring before the release. *[Hint: see Kinetic Energy P34]*

20.5 Two trolleys are moving in opposite directions along a smooth surface. The trolleys collide and stick together.

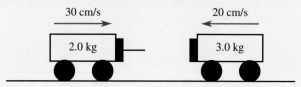

(a) What is the total momentum of the trolleys before and after the collision?

(b) What is the trolleys' combined velocity after the collision?

20.6 Complete the following table for two objects A and B, travelling together until pushed apart by explosions. 'm_A' means 'mass of A'.

m_A (kg)	m_B (kg)	Initial combined velocity (m/s)	Final v_A (m/s)	Final v_B (m/s)
2.5	2.5	0.0	(a)	+2.0
2.5	5.0	0.0	−6.8	(b)
2.5	7.5	5.0	−4.0	(c)
9.0	(d)	−4.0	−5.0	+6.0

20.7 A 10 g bullet (0.010 kg) is fired at 250 m/s Eastwards towards a 10 kg sandbag.

    (a) Calculate the momentum of the bullet.

    (b) What is the total momentum before the collision?

    The bullet enters the sandbag and stops inside it:

    (c) What is the total momentum now?

    (d) Calculate the speed of the sandbag.

20.8 A 70 kg astronaut has a 20 kg backpack, and is stranded, stationary, in space 30 m to the West of her spacecraft. To get back to safety, she hurls the backpack at a speed of 4.2 m/s.

    (a) Which way does she need to throw the backpack?

    (b) What is the total momentum before she throws it?

    (c) What is the momentum of the backpack after throwing?

    (d) What will the astronaut's momentum be after she has thrown the backpack?

    (e) What is the astronaut's velocity after she has thrown the backpack?

    (f) How much time does it take her to get back to the spacecraft?

20.9 The exhaust from a rocket on a test rig leaves the engine at 2 800 m/s. How many kilograms of propellant (fuel and oxidizer) need to be burnt every second to provide a force of 3.5×10^8 N?

20.10 A conveyor belt is used to move coal along a horizontal shaft in a coal mine. How much force needs to be applied horizontally to the belt to keep it moving at 1.2 m/s if 40 kg of coal is dropped onto it every second? Assume that the coal has no horizontal velocity before it touches the belt and the belt's turning mechanism is well lubricated.

$^{23}/_{30}$

21   Motion with Constant Acceleration

The equations we will develop and practise here can be used in any situation where the acceleration does not change. As long as drag forces are small enough to be ignored, this includes:

- anything falling freely;

- anything speeding up because an engine is providing a steady force on it;

- anything slowing down because brakes are providing a fixed force.

We start with three principles:

1. Displacement $=$ average velocity $\times$ time

2. Velocity change $=$ acceleration $\times$ time

3. If the acceleration is constant, then the velocity will rise steadily. This means that the average velocity will be half way between the starting and final velocities (it will be the mean of the starting and final velocities).

In this book, we use five letters to represent the quantities.

Letter	Quantity	Unit
s	Displacement	m
u	Starting velocity	m/s
v	Final velocity	m/s
a	Acceleration	m/s^2
t	Time taken	s

We can write our three principles as equations using these letters. Firstly, the third principle means that average velocity $= 1/2\,(u + v)$.

1. $s = \left(\dfrac{u + v}{2}\right) t$ 　　　　　　　　　2. $v - u = at$

Now rearrange equation (2) to make v the subject; and then substitute this into equation (1). This gives

$$v = u + at \quad \text{so} \quad s = \left(\frac{u + u + at}{2}\right)t = \frac{2ut + at^2}{2} = ut + \tfrac{1}{2}at^2$$

Next, rearrange equation (2) to make t the subject; and then substitute this into equation (1). Finally, rearrange it to make v^2 the subject. This gives

$$t = \frac{v - u}{a} \quad \text{so} \quad s = \left(\frac{u + v}{2}\right) \times \left(\frac{v - u}{a}\right) = \frac{(u + v)(v - u)}{2a} = \frac{v^2 - u^2}{2a}$$

$$\text{so} \quad v^2 = u^2 + 2as$$

Let's look at our four equations, often given in examination formula sheets.

$$v = u + at \qquad \text{has no } s \qquad\qquad v^2 = u^2 + 2as \qquad \text{has no } t$$

$$s = \left(\frac{u + v}{2}\right)t \qquad \text{has no } a \qquad\qquad s = ut + \tfrac{1}{2}at^2 \qquad \text{has no } v$$

Example 1 – An aeroplane requires a speed of 26 m/s to take off. If its acceleration is 2.3 m/s^2, how much runway does it 'use up' before it lifts off? Assume it starts at rest.

Using basic principles:
Time = velocity gained / acceleration = 26 m/s $\div$ 2.3 m/s^2 = 11.3 s
Average velocity = $\tfrac{1}{2}$ (0.0 m/s + 26 m/s) = 13 m/s
Displacement = av. velocity $\times$ time = 13 m/s $\times$ 11.3 s = 150 m (2 sf)

Using the equations:

$u = 0$ m/s $\qquad\qquad v = 26$ m/s $\qquad\qquad a = 2.3$ m/s^2 $\qquad$ we want to know s

We use the equation with no t as we don't know t.
$v^2 = u^2 + 2as$, so $26^2 = 0^2 + 2 \times 2.3 \times s$
so $676 = 4.6 \times s$, so $s = 676/4.6 = 150$ m (2 sf)

Example 2 – How much time does it take a ball to fall 30 cm if it is accelerating downwards at 10 m/s² after being dropped?

NB: 'dropped' means it isn't moving to start with, so $u = 0$.

Using the equations:

$s = 0.30$ m $u = 0$ m/s $a = 10$ m/s²   we want to know t

We use the equation with no v as we don't know v:
$s = ut + 1/2\,at^2$, so $0.30 = 0t + 1/2\,10t^2$, so $0.3 = 5t^2$
$t^2 = 0.3/5 = 0.06$ so $t = \sqrt{0.06} = 0.24$ s

Using basic principles (where we use t to represent the time):
Velocity change = acceleration × time = $10t$
Final velocity = initial velocity + velocity change = $0 + 10t = 10t$
Average velocity = $1/2\,(0 + 10t) = 5t$
Displacement = average velocity × time = $5t \times t = 5t^2 = 0.30$
so $t^2 = 0.30/5 = 0.06$ so $t = \sqrt{0.06} = 0.24$ s.

21.1   Complete the table, where each row is a separate question.

s (m)	u (m/s)	v (m/s)	a (m/s²)	t (s)
	2.0	(a)	3.0	6.0
(b)	2.0		3.0	6.0
(c)	0.0		10	0.20
(d)	0.0		10	0.40
16	1.5	(e)		10
0.82	14	0.0		(f)
	31	0.0	−6.7	(g)

21.2   An old £5 note is 135 mm long. A friend has a crisp £5 note, and holds the bottom of the note in line with (and between) your thumb and index finger. She drops it, and if you grab it without moving your hand downwards, you are allowed to keep it. How quickly do you have to react to win your prize?

21.3 The Highway Code assumes that a car with its brakes on fully has an acceleration of -6.7 m/s^2. Calculate the

(a) time taken to stop a car from 30 mph (13.4 m/s);

(b) distance taken to stop a car at 30 mph;

(c) time taken to stop a car from 70 mph (31 m/s);

(d) distance taken to stop a car from 70 mph.

21.4 You throw a cricket ball up into the air at 10 m/s. [Hint: if you take $u = 10$ m/s then $a = -10$ m/s^2 as the acceleration is in the opposite direction to the initial velocity.]

(a) How much time elapses before it reaches the highest point of its motion? [Hint: at the top, v=0.]

(b) How high does it go?

21.5 If there were no air resistance, how much time would it take for a dropped parcel to fall 2 000 m?

21.6 What is the deceleration of a train which takes 2.3 km to stop from a speed of 67 m/s?

21.7 How much time does it take to stop an oil tanker if its speed is 8.0 m/s to start with, and the stopping distance is 5.0 miles? One mile is about 1 600 m.

21.8 The Eiffel Tower is 300 m high. A coin is dropped from the top; how fast is it going when it hits the pavement? Assume no air resistance.

21.9 How fast would you have to shoot a scientific instrument upwards if you wanted it to rise 200 km above the Earth's surface ignoring air resistance?

21.10 The acceleration of dropped objects on the Moon is 1.6 m/s^2. How long does it take a feather to fall 0.70 m? [There is no air resistance!]

Electricity

Electric charge is a property of matter. There are two types of electric charge, which are conventionally labelled positive and negative. Two objects that have the same charge exert repulsive forces on each other. Two objects that have opposite charges exert attractive forces on each other.

Electric charge is measured in coulombs (C) and has the symbol Q.

Electric charge is quantised, which means any object can only have an integer multiple of a certain value of charge. The smallest value of the magnitude of charge an object can have is equal to the magnitude of the charge of an electron, which is approximately 1.60×10^{-19} C. Electrons are negatively charged. If a neutral object loses electrons, it becomes more positively charged. If a neutral object gains electrons, it becomes more negatively charged.

Current is the rate of flow of charge. Current can be caused by the flow of electrons, ions or other charged particles. When we give a direction, we give the direction a positive ion would go. Accordingly electrons, which are negatively charged, actually move in the opposite direction to the arrow on the circuit diagram.

The equation relating electric charge, current and time is:

$$\text{electric charge} = \text{electric current} \times \text{time} \quad Q = It$$

In an electric circuit, electric current flows from the positive terminal of a power supply to the negative terminal or ground, or from the ground to a negative terminal.

Electric current is measured in amperes (A) and has the symbol I.

Example 1 – If a charge of 105 C flows in 15 s, calculate the current.
Current = charge / time = 105 C / 15 s = 7.0 A (2sf)

Number of electrons used to carry charge = charge / charge of one electron

Number of electrons which flow past a point in one second = charge flow in one second / charge of one electron = current / charge of one electron

---

Example 2 – Calculate the number of electrons which are required to carry a charge of 105 C.
Number of electrons = charge / charge of one electron =
$105 \, C \, / \, 1.6 \times 10^{-19} \, C = 6.6 \times 10^{20}$ (2sf)

Assume here electrons have a charge of -1.60×10^{-19} C:

22.1 A 3.00 A appliance has 360 C of charge flow through it.

   (a) How long was the appliance operating?

   (b) How many electrons passed through the appliance in this time?

22.2 -1.00 coulomb is the charge of how many electrons?

22.3 If two electrons are removed from an atom, what is the charge of the resulting cation (a positively charged ion)?

22.4 Work out the missing measurements from the following table, where each row is a separate question.

Charge	Current	Time
(a)	2.50 A	900 ms
350 mC	(b)	20.0 s
900 µC	18.0 µA	(c)
(d)	3.0 A	10 s
(e)	0.60 A	60 s
60 C	1.9 A	(f)
1 260 C	0.50 A	(g)
110 C	(h)	25 s
3.3×10^4 C	(i)	6.0×10^3 s

22.5 An appliance draws a current of 9.00 A.

    (a) How much charge flows in 5.0 minutes of operation?

    (b) How many electrons flow in 5.0 minutes?

22.6 5.40×10^{21} electrons flow through another appliance in 3.0 minutes.

    (a) How much charge flows through this appliance?

    (b) What is the current drawn by this appliance?

22.7 A car headlamp bulb draws 1.9 A from the battery. How much charge flows through the bulb in 5.0 minutes?

22.8 How much charge flows through a lamp in 2.0 minutes if it carries a steady current of 0.30 A?

22.9 What current flows through a resistor if 1 260 C of charge moves through it in 5.0 minutes?

22.10 How long would a charge of 600 C take to move round a circuit at a steady rate of 12 C/s?

Additional Charge and Current Questions – with more on-line

22.11 With a steady current of 5.0 C/s, how long would it take to move 200 C of charge through a resistor?

22.12 In an electric circuit, an ammeter, in series with a 2 700 Ω resistor, reads 4.5 mA. How many coulombs of charge pass through the resistor in a time of 5.0 minutes?

22.13 How long has a lamp been switched on if it draws 2.5 A from its power supply and 75 C of charge has passed through it?

22.14 In part of an electronic circuit, a component is charged in 9.0 s by an average current of 5.0 mA. How much charge is delivered to the component in this time?

22.15 Car batteries are rated in units called amp-hours. This is the time in hours for which the battery can supply a 1 A current. What quantity is measured in this unit?

23   Current and Voltage - Circuit Rules

Current is the rate of flow of electric charge. Current is not used up in a circuit; at all points in a series circuit, current has the same value.

If a circuit has a branch, the current flowing into the junction must equal the current flowing out of it.

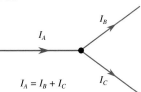

In the diagram above, the value of Current A is equal to the sum of the values of Current B and Current C.

Voltage is also known as potential difference. The voltage across a component is the work done per unit charge in driving the charge through the component. In a circuit loop, the sum of the voltages across the power supplies is always equal to the sum of the voltages across the rest of the components; see the left figure below.

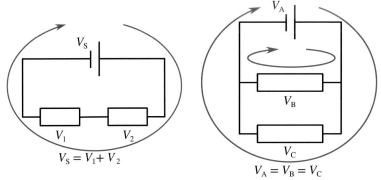

In the right diagram above, the value of the voltage across the cell is equal to the value of the voltage across the top resistor (the top loop of the circuit) and also to the value of the voltage across the bottom resistor (the bottom loop of the circuit).

This means components in parallel have equal voltage, and components in series divide the available voltage between them.

	Components in series . . .	Components in parallel . . .
Current	have the same current as each other	divide the current between them
Voltage	divide the voltage between them	have the same voltage as each other

Assume below perfect voltmeters and ammeters. Perfect voltmeters carry no current. Perfect ammeters have zero potential difference across them.

23.1 In the circuits below, state the current at positions P and Q.

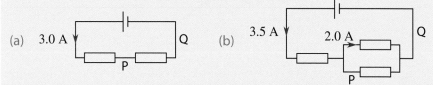

(a)   3.0 A   Q   (b)   3.5 A   2.0 A   Q
                              P   P

23.2 In the circuits below, state the potential difference across (a) resistor R, and (b) R_1 and R_2.

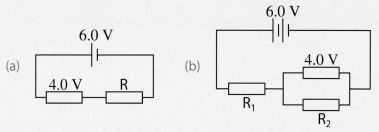

(a)   6.0 V   (b)   6.0 V

      4.0 V   R   4.0 V
                                                          R_1
                                                                        R_2

23.3 What are the readings on the voltmeters V_1, V_2 and V_3 below?

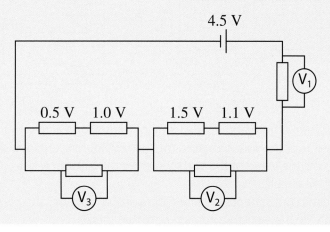

4.5 V

                                                                V_1

0.5 V   1.0 V   1.5 V   1.1 V

                V_3 V_2

23.4 In the circuit below, what are the readings on the ammeters A_1, A_2 and A_3?

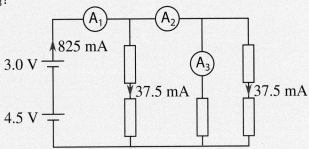

23.5 In the circuit below, what are the readings on the ammeters A_1 and A_2 and the voltmeters V_1 and V_2?

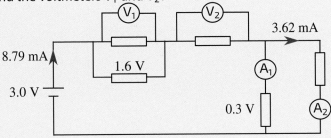

23.6 In the circuit below, what are the missing voltages; A and B?

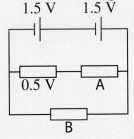

23.7 In the circuit below, what are the missing currents; X, Y and Z?

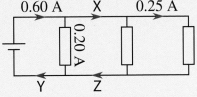

23.8 In the circuit below, what are the missing currents (F and G) and
      voltages (H, J, K)?

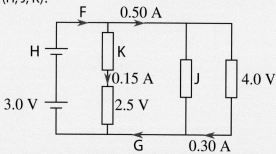

23.9 What are the readings on A_1 and A_2, and on V_1 and V_2 below?

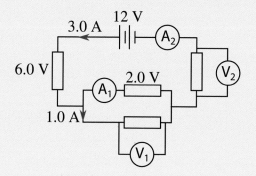

23.10 What are the readings on A_1, A_2 and A_3, and on V_1, V_2 and V_3 below?

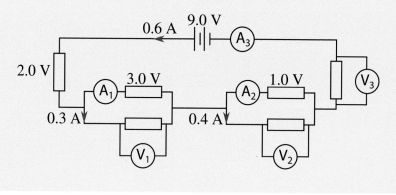

24   Resistance

Resistance measures how difficult it is for electric current to pass through a component (or through an object) for an applied voltage. A resistor is a circuit component that dissipates energy thermally when work is done in driving a current through it.

So an iron nail has less resistance than a plastic pen.

Resistance is measured in ohms (Ω – the upper case Greek letter "omega").

Formula:

resistance (Ω) = voltage across component (V)/current through it (A)

$$R = V/I, \text{ so}$$
$$V = IR$$

24.1 Complete the table, where each row is a separate question:

Voltage (V)	Current (A)	Resistance (Ω)
6.0	1.98	(a)
240	0.020	(b)
(c)	0.050	200
415	(d)	200

24.2 What is the resistance of the heating element in an electric oven which carries a current of 10 A when connected to 230 V mains?

24.3 What resistance is needed if you wish to have a 10 mA current and the supply voltage is 20 V?

Notice from your answer from Q24.3 that the equation $V = IR$ also works if you measure I in mA and R in kΩ. This is useful as the mA and the kΩ are more convenient units in electronics than the amp and ohm.

24.4 A car headlamp bulb has a filament resistance of 6.0 Ω. If the car battery is 13.2 volts, how much current does the bulb take from the battery when lit?

24.5 The heater of a toilet hand drier uses 9.0 amps from a 240 volt mains supply. What is the resistance of the heater's element?

24.6 (a) What is the voltage across a 0.33 kΩ resistor carrying 20 mA?

(b) A current of 3.33 mA flows through a 2.7 kΩ resistor. Calculate the potential difference across the resistor.

(c) In part of an electric circuit, the potential difference across a 12 Ω resistor is 16 V. What is the current through the resistor?

24.7 Some resistors are labelled with a red band. This shows that their true resistance will be within 2.0% of the value shown on the side. What is the largest current you would expect through a 15 kΩ, 2.0% resistor when connected to a 5.0 V supply?

24.8 When both of a car's red tail lamps are lit, this 'draws' a current of 0.83 A from the battery. Given that a car battery has a voltage of 12 V, what is the resistance of each lamp? The lamps are wired in parallel.

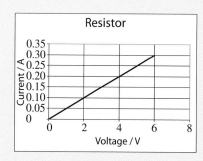
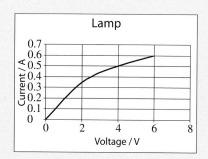

24.9 Study the two graphs above, showing the current passing through a resistor and lamp for different supply voltages.

(a) What is the resistance of the resistor when it is connected to a 6 V supply?

(b) What is the resistance of the resistor when it is carrying 0.20 A?

(c) What current would you expect to flow through the resistor if the voltage across it were 10 V?

(d) What is the resistance of the lamp when it is connected to a 4 V supply?

(e) If you wired the resistor and the lamp in parallel and connected them to a 6 V supply, how much current would be 'taken' in total from the supply?

(f) *[Harder]* If you wired the resistor and the lamp in series, and connected this to a 14 V supply

    i. what would the current be?

    ii. what would be the voltage across the lamp?

24.10 *[Harder]* When a light emitting diode (LED) is connected into an electric circuit, it is wired in series with a resistor to ensure it doesn't get damaged by taking too much current. The voltage across a red LED is 1.5 V when it is lit. If you wish to supply it using a 9.0 V battery, what resistance of resistor is needed if a 30 mA current is needed?

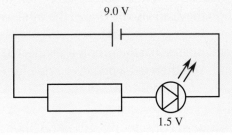

9.0 V

1.5 V

25   Characteristics

Component characteristic graphs can be used to predict the amount of current drawn by an electrical component when a certain potential difference is across it. With these two values, the resistance of the component can be calculated using the equation,

$$\text{resistance} = \text{voltage}/\text{current} \qquad R = V/I$$

A voltage-current graph that is a straight line through the origin shows that the resistance of the component is independent of the potential difference across it, or the current flowing through it.

A graph with a curved line shows that the resistance depends on the potential difference applied across it; the resistance does not have a constant value.

Characteristic graphs are typically drawn with the current (I) on the y-axis (the vertical axis) and the potential difference (V) on the x-axis (the horizontal axis), although they can also be drawn the other way around.

A negative value for V means that the supply is connected to the component the other way round. You then get a negative value for I meaning that the current is now flowing the opposite way through it.

25.1 For these questions consider the following graph. When reading values from the axes, round to the nearest integer.

(a) What current is drawn at a voltage of 2 V?

(b) What current is drawn at a voltage of 6 V?

(c) What is the resistance at 1 V?

(d) What is the resistance at 10 A?

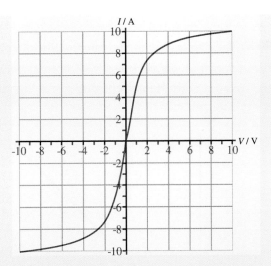

25.2 The following characteristic graph is for a typical diode.

(a) What is the resistance at 1 V?

(b) Does the resistance change above 1 V? Explain your answer.

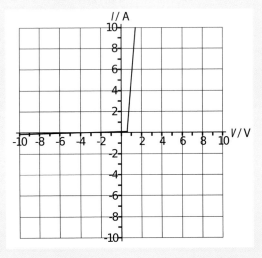

25.3 An ohmic device is one where the resistance does not depend on voltage for fixed physical conditions (e.g. temperature). What two

features would you expect to see in a characteristic graph of an ohmic device?

25.4 The following graph shows how the potential difference across a conductor varies with the current through it. Calculate the resistance of the conductor.

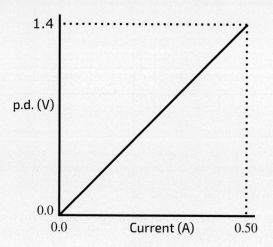

25.5 The graph below shows how the potential difference across a conductor varies with the current through it. How does the resistance of the conductor change with increasing current?

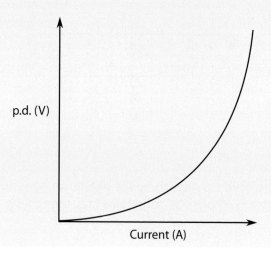

26   Power Calculations

Power is the rate at which energy is transferred, or the rate at which work is done.
It is calculated using the equation:

$$\text{power} = \text{work done}/\text{time} \qquad P = \frac{W}{t}$$

$$\text{or} \quad \text{power} = \text{energy transferred}/\text{time} \qquad P = \frac{E}{t}$$

The unit of power is the watt (W).

$$1 \text{ watt} = 1 \text{ joule per second}$$

$$1 \text{ W} = 1 \text{ J/s}$$

26.1   Calculate the missing quantities in the table below.

Power	Energy transferred	Time
(a)	15.5 J	25.0 s
250 W	(b)	60.0 s
150 W	5 250 J	(c)
(d)	105 kJ	3 min 40 s
250 mW	(e)	100 ms
0.125 kW	96.0 µJ	(f)

Electrical power
Potential difference, or voltage, across a component is the amount of work done per unit charge moving through that component, i.e.

$$\text{potential difference} = \text{work done}/\text{charge} \qquad V = \frac{W}{Q}$$

Electric current is the amount of charge that flows past a point per second, i.e.

$$\text{current} = \text{charge}/\text{time} \qquad I = \frac{Q}{t}$$

Multiplying these quantities together gives:

$$I \times V = \left(\frac{Q}{t}\right) \times \left(\frac{W}{Q}\right)$$

The Qs cancel, giving:

$$I \times V = \frac{W}{t} = P$$

which is equal to power (first equation on the page). So, the equation for electrical power is:

$$\text{power} = \text{current} \times \text{potential difference} \qquad P = I \times V$$

26.2 Calculate the missing quantities in the table below.

Power	Current	Potential difference
(a)	0.250 A	1.50 V
22.2 W	(b)	6.00 V
1 200 W	80.0 A	(c)
(d)	68.0 µA	5.00 kV
8.16 kW	(e)	8.50 kV
4.05 MW	54.0 mA	(f)
(g)	5.0 A	12 V
2.64 × 10⁶ W	(h)	132 000 V
0.366 W	0.060 A	(i)
(j)	10 A	230 V
1 000 W	(k)	230 V
72 W	6.0 A	(l)

26.3 How much current does a 2.0 kW electric fire draw from the 230 V mains?

26.4 What is the power rating of a lamp which draws 0.26 A from the 230 V mains?

26.5 A torch bulb has 2.5 V, 0.18 A stamped on it. What is its power rating?

26.6 What is the potential difference across a heater which develops power of 42 W when a current of 3.5 A flows through it?

26.7 The power of the heater element of a toilet hand dryer is 2 100 W. It operates from the 230 V mains. Calculate the current drawn from the mains.

26.8 What current is carried in the element of a 2.4 kW kettle connected to the 230 V mains?

26.9 An MRI scanner has a peak power of 35.0 kW. It is connected to a power supply at 415 V.

(a) What is the peak current drawn by the MRI scanner?

(b) If a scan takes 30 minutes to complete, how much charge has flowed through the scanner in total? Assume the current is constant and equal to Part (a)

26.10 On building sites, 115 V mains is used to reduce the risk of electric shock. A drill made for normal household (230 V) use requires a current of 5.60 A. The manufacturer makes a model of the same power rating for use on building sites. What current will the builders' version need?

$^{21}/_{27}$

27   Resistance and Power

Equations:

$$\text{voltage} = \text{current} \times \text{resistance} \qquad V = IR$$
$$\text{power} = \text{current} \times \text{voltage} \qquad P = IV$$

Example 1 – Calculate the power dissipated in a $6.0\ \Omega$ resistor carrying 3.5 A.

Voltage $= IR = 3.5\,\text{A} \times 6.0\,\Omega = 21\,\text{V}$

Power $= IV = 3.5\,\text{A} \times 21\,\text{V} = 73.5\,\text{W} = 74\,\text{W}$ (2 sf)

Eliminating V, we have:

$$\begin{aligned} P &= I \times V = I \times (IR) = I^2 R: \quad \text{rearranging gives} \\ I^2 &= P/R \quad \text{and} \quad R = P/I^2. \end{aligned}$$

Example 2 – Calculate the resistance of a heater if it needs to carry 13 A when dissipating 3 100 W.

$R = P/I^2$, so $R = 3100/169 = 18\ \Omega$ (2 sf)

Eliminating I, we have:

$$\begin{aligned} P &= I \times V = (V/R) \times V = V^2/R: \quad \text{rearranging gives} \\ V^2 &= PR \quad \text{and} \quad R = V^2/P. \end{aligned}$$

Example 3 – Calculate the power dissipated when a $200\ \Omega$ resistor is connected to a 240 V supply.

$P = V^2/R = 240^2/200 = 290\,\text{W}$ (2 sf)

Example 4 – Calculate the resistance of a 50 W light bulb connected to a 12 V supply.

$R = 12^2/50 = 2.9\ \Omega$ (2 sf)

27.1 Use a two stage method, as in example 1, where you need to first calculate the unknown voltages or the unknown currents: Calculate the power when

(a) 3.2 A flows through an 18 Ω resistor;

(b) 32 A flows through a 2.0 Ω wire;

(c) an 18 Ω resistor is connected to a 24 V supply;

(d) a 2.0 Ω wire is put across the terminals of a 240 V supply.

27.2 Complete the table, where each row is a separate situation.

Voltage	Current	Resistance	Power
9.0 V	(a)	300 Ω	(b)
240 V	13 A	(c)	(d)
240 V	(e)	25 Ω	(f)
(g)	100 A	3.0 Ω	(h)
240 V	(i)	(j)	2 500 W
240 V	(k)	(l)	60 W
23 kV	(m)	(n)	23 MW
9.0 V	(o)	22 kΩ	(p)
(q)	30 mA	(r)	0.75 W

27.3 What voltage is needed if 3.0 W of power is going to be dissipated in a 4.5 Ω resistor?

27.4 Some of the power coming into houses is wasted by the wires carrying the current due to their resistance.

(a) For a particular house, the wire which supplies it has a resistance of 1.5 Ω. If the current is 83 A, what is the power wastage in the supply wire?

(b) For a different house, the rules say that no more than 6.0 V may be 'dropped' across the supply wire. What power wastage does this correspond to if the wire has a resistance of 2.5 Ω?

27.5 What is the resistance of a 1.2 kW light bulb operating on a voltage of 115 V?

27.6 Old lamp dimmers were variable resistors wired in series with the light bulb. Suppose you put a 25 Ω resistor in series with a lamp such that the voltage across the lamp is only half of the 230 V supply voltage. What is the power dissipated by the resistor?

27.7 The National Grid operates at voltages of up to 400 kV. A generator has an output power of 68 MW at 400 kV. If the wire supplying customers has a resistance of 6.5 Ω, calculate

(a) the current in the wire;

(b) the voltage 'dropped' along the wire, and;

(c) the power 'wasted' in the wire.

27.8 A resistor carries a current of 2.0 A.

(a) If its resistance is 50 Ω, what is the power developed in it?

(b) If its resistance is 48 Ω, what is the power developed in it?

27.9 What is the resistance of a resistor which develops 1 000 watts of power when 10 amps flows through it?

27.10 In the UK, mains voltage is 230 V.

(a) Calculate the power of a hair dryer element which is designed to operate from the mains and has an element of resistance 57.6 Ω.

(b) What electric power is used by a light bulb which has a filament resistance of 1 440 Ω and works on mains voltage?

28 E-M Induction and Generators ♡

When a wire is moved in a magnetic field, a voltage is induced, providing that the wire is moved so that it cuts the magnetic field lines.

You can reverse the direction of the voltage by

- moving the wire in the opposite direction, or by
- reversing the direction of the magnetic field.

You can increase the voltage by

- moving the wire more quickly, or by
- using a stronger magnetic field.

When a magnet is moved into a coil of wire, a voltage is induced. You can make it larger by

- moving the magnet more quickly,
- using a stronger magnet, or by
- using a coil with more turns of wire on it.

You can reverse the direction of the voltage by

- moving the magnet in the opposite direction, or by
- using a magnet magnetized the other way.

In fact, the voltage is proportional to the magnetic field strength, the speed of movement across the field lines, and the length of wire in the field. This means that, if there is no relative motion (the magnet is stationary in the coil, or the wire is stationary in the magnetic field) no voltage is induced – no matter how strong the field is (providing the field strength is not changing). Generators are mechanical devices (with wires and a magnetic field) so that mechanical work done on the generator (usually to turn it) enables the generator to do electrical work on the circuit (and light bulbs). The work done on the generator or by moving a wire across the field, does not reduce the energy stored in the magnetic field. Any energy transferred comes from the work done in moving the wire and not from the magnetic field itself.

28.1 A long wire, connected to a centre-zero galvanometer, is moved
      downwards perpendicular to a magnetic field. The field is between
      two permanent magnets, with opposite poles facing each other.
      While the wire is moving, the galvanometer needle moves to the
      right.

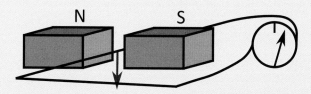

      (a) How would the pointer of the galvanometer move if the wire
      was moved up through the magnetic field?

      (b) What would be the induced current in the conductor if it was
      held stationary in the centre of the magnetic field?

      (c) State three ways of increasing the deflection of the pointer on
      the galvanometer when the wire is moved through the magnetic
      field.

28.2 A pupil is investigating the effects of electromagnetic induction.

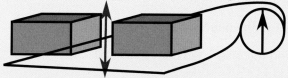

      They move a conducting wire up and down parallel to the faces
      of the two permanent magnets, expecting the pointer on the gal-
      vanometer to show a deflection. However, the pointer does not
      move. The meter is working and sensitive enough for the experi-
      ment. There are no breaks in the wires. What must be wrong?

28.3 Suppose +1.5 mV is induced when a wire is moved between the
      poles of a large permanent magnet at a speed of 0.20 m/s from left
      to right. What voltage would you expect when: [Hint: for an explan-
      ation of solving proportionality questions turn to P16]

      (a) the speed is increased to 0.40 m/s?

(b) the speed is 0.3 m/s, but the wire is moved from right to left?

(c) the original experiment is repeated with a magnet four times as strong?

28.4 Suppose $+2.7$ V is induced when the North pole of a magnet is inserted into a 200-turn coil at a speed of 1.5 m/s from above. What voltage would you expect if

(a) the coil had 450 turns on it?

(b) the magnet were moved into the 200-turn coil at 8.4 m/s?

(c) the South pole of the magnet were moved into the 200-turn coil at 4.5 m/s?

(d) the North pole of a magnet with $1\,000\times$ the strength was held still inside the coil?

28.5 A simple electrical generator is made by mounting a magnet (shaded) in a coil of wire, with the magnet continuously rotating slowly about a horizontal axis. It is connected to a meter which can read positive and negative voltages. One quarter of a turn later from the position shown on the diagram, the meter shows a positive voltage of 1.5 V.

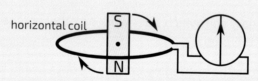

(a) What will the meter show one half of a turn later still? (i.e. three quarters of a turn on from the position shown in the diagram?)

(b) What will the meter show when the magnet is back in the position shown in the figure?

(c) What two things would happen if you turned the magnet twice as quickly?

(d) How could you modify the design so that larger voltages were induced without needing to change the speed of the rotation?

$^{12}/_{15}$

29   Transformers

While a very strong magnet held stationary inside a coil will not induce a voltage (there is no relative motion), if the magnetic field gets stronger a voltage will be induced. This is because the increase in magnetic field at the coil could have been caused by an ordinary magnet moving closer. Permanent magnets cannot change strength readily, but you can change the strength of an electromagnet if you change the current flowing in it.

This is the principle of the transformer. Transformers only work on alternating current (a.c.). The current in the primary coil causes it to become an electromagnet. The continually changing current produces a continually changing magnetic field in an iron core. This in turn induces a continually changing voltage in the nearby secondary coil wound round the iron core. A transformer won't work on direct current (d.c.) because a stationary magnet will only produce a steady magnetic field - and steady, stationary magnetic fields do not induce voltages. A transformer does not change the frequency of the alternating current.

Transformers have two coils

- the primary coil, connected to an a.c. supply of known voltage, and

- the secondary coil, which does work on other components using energy from the primary.

The voltage across the secondary coil, V_s, is not usually the same as the primary coil's supply voltage, V_p. It could be greater (a step-up transformer) if the number of turns on the secondary is greater, $N_s > N_p$, or less if the number of turns on the secondary is fewer, $N_s < N_p$.

$$\text{secondary (a.c.) voltage} = \text{primary (a.c.) voltage} \times \frac{\text{no. of turns on secondary}}{\text{no. of turns on primary}}$$

$$\frac{V_s}{V_p} = \frac{N_s}{N_p} \quad \text{or} \quad \frac{N_p}{V_p} = \frac{N_s}{V_s}$$

This means that the number of 'turns per volt' is the same on both coils.

Example 1 – A transformer has an input voltage of 240 V a.c. and output of 48 V. If there are 3 000 turns on the primary coil, how many are there on the secondary?

$V_s/V_p = N_s/N_p$, so $48/240 = N_s/3000$.
Thus $0.2 = N_s/3000$, so $N_s = 0.2 \times 3000 = 600$ turns.
Or, you could solve it like this: the primary coil has $3000/240 = 12.5$ turns/volt
So the secondary must have $48 \times 12.5 = 600$ turns.

29.1 Complete the table below. Each row is a separate question.

a.c. Voltage (V)		No. of turns on coil		Step up or down?
Primary	Secondary	Primary	Secondary	
240	(a)	2 000	200	(b)
11 000	240	(c)	600	(d)
23 000	230 000	(e)	1 000	(f)
240	12	(g)	300	(h)
240	4.96	1 500	(i)	(j)

29.2 A doorbell for a house works from 8.0 V a.c. To operate the bell from the 240 V mains supply, a transformer can be used.

(a) How many turns would be in the primary winding for each turn in the secondary winding?

(b) Would the transformer be a step-up or a step-down type?

29.3 To produce an output of 48 V a.c. from an input of 240 V a.c., how many turns would be required in the primary winding if there were 100 turns in the secondary?

29.4 The input voltage to a step-down transformer is 240 V a.c. at a frequency of 50 Hz. The primary winding has 6 000 turns and the secondary 300 turns.

(a) What is the voltage output?

(b) What is its output frequency?

29.5   A step-up transformer has 500 turns in the primary coil and 10 000 turns in the secondary coil. A voltage of 250 V a.c. is applied to the primary at 50 Hz.

(a) What is the voltage of the output at the secondary?

(b) What is the frequency of the output at the secondary?

29.6   A 12 volt car battery is placed across the primary coil of a 1:20 step-up transformer. What is the output voltage across the secondary?

29.7   These questions should show you why we bother with transformers on our electrical distribution system. Calculate:

(a) The current needed to distribute the 2 000 MW generated at 22 kV at a large power station.

(b) The current needed to carry 2 000 MW if the voltage is 400 kV.

The cables used have a total resistance of 9.0 Ω.

(c) Calculate the power wasted in heating the wire if the current from 22 kV flows in the wire.

(d) Calculate the power wasted if the current carrying 2 000 MW at 400 kV passes through the wire.

29.8   A computer power supply unit can be switched to work on European (230 V a.c.) or United States (115 V a.c.) mains. In the European setting, there are 2000 turns on the primary coil. When switched into United States mode, how many turns are now on the primary?

29.9   If 0.45 A were flowing in the primary of part (a) in the above table, what would the current in the secondary be if the transformer were 100 % efficient? [Hint: Power = Current × Voltage.]

29.10   The power station generator of 29.1(e) produces a current of 22 kA. If the transformer is 97 % efficient, what will the output current onto the 230 kV national grid be? NB - Your answer to Q29.1(e) does not need changing.

Energy

Energy analysis determines whether some processes are possible. It involves calculating the amounts of energy stored in different places and in different ways. An energy analysis is one of many ways of examining physical processes. If we want to explain how a microphone works, there is lots to discuss but little benefit from mentioning energy. However, if we want to know how much fuel is needed to lift a satellite into space, then we must perform calculations based on energy.

You will calculate energy as it is stored: thermally, gravitationally, elastically, as kinetic energy, and as nuclear energy. Whilst the energy stored in these different ways may differ before and after a physical change, the **total** energy is the same.

30   Thermal Energy

Hot objects (or substances) store energy thermally.

The energy is associated with the random, thermal motion of a substance's atoms or molecules. The physical processes of conduction, convection and radiation can result in increases or decreases of a thermal energy store.

If two objects at different temperatures are in contact with each other, then energy is exchanged between their particles. After some time the thermal energy store of the hotter object will have decreased (and its temperature will have decreased) and the thermal energy of the cooler object will have increased (and its temperature will have increased). This thermal process is called conduction. When the objects reach the same temperature, we say they are in thermal equilibrium and the thermal energy of each will then remain constant.

Hot objects emit electromagnetic radiation. After a period of time, the thermal energy store associated with a hot object will have decreased (and the temperature of the hot object will be lower than it was).

Thermal energy is measure in joules (J).

Heating involves the transfer of thermal energy from a warmer (higher temperature) object to a cooler (lower temperature) object.

The amount of thermal energy required to increase the temperature of an object (of a certain substance) by $1\,°C$ is called the heat capacity of that object. The heat capacity per kilogram is called the specific heat capacity (of that substance). Therefore,

change in thermal energy $=$ mass $\times$ specific heat capacity $\times$ change in temp.

$$\Delta Q = mc\Delta T$$

Specific heat capacity is measured in J/(kg $°C$) or the equivalent unit J/(kg K).

The specific heat capacity of pure water is $4\,200$ J/(kg $°C$).

30.1 Work out the missing measurements from the following table, where each row is a separate question.

ΔQ	m	c (J/(kg $°C$))	ΔT ($°C$)
(a)	2.50 kg	800	30.0
$5\,625$ J	(b)	750	15.0
34.125 kJ	250 g	(c)	65.0
69.0 kJ	1.20 kg	250	(d)
(e)	2.0 kg	$4\,200$	10
$90\,000$ J	(f)	450	10
2.1×10^5 J	10 kg	(g)	5.0
$8\,000$ J	1.0 kg	400	(h)
(i)	0.50 kg	$4\,200$	40
$10\,500$ J	(j)	$2\,100$	20
$67\,500$ J	5.0 kg	(k)	30
2.5×10^4 J	0.50 kg	$1\,000$	(l)

30.2 What is the change in thermal energy of 1.00 kg of water that is raised from 20.0 $°C$ to its boiling point? (Assume the system is well insulated.)

30.3 How much energy is required to raise the temperature of a 200 g block of ice from $-10.0\,°C$ to $0.0\,°C$? (The specific heat capacity of ice is 2 100 J/(kg °C). Assume the system is well insulated).

30.4 The specific heat capacity of ice is 2 100 J/(kg K). A 1.8 kg block of ice, removed from a freezer at a temperature of $-18\,°C$, is placed in a fridge which has a temperature of $0.0\,°C$. After a few hours, the ice has warmed up to the fridge temperature. What is the change in the stored thermal energy of the block?

30.5 The specific heat capacity of air is 1000 J/(kg K).

(a) How much energy would be needed to raise the temperature of the air in a room by $5.0\,°C$, if the room measures 4.0 m $\times$ 4.0 m $\times$ 3.0 m? (Take the density of air $= 1.0$ kg/m³). Assume that the room has no furniture and that the walls gain no thermal energy.

(b) How long would a 1.0 kW convection heater take to heat the air?

30.6 A carpet cleaning machine holds 40 litres of water. It is filled with water at $15\,°C$ and the water is then heated by a 3.0 kW element to a temperature of $70\,°C$.

(a) How much energy is added to the water? [Mass of 1.0 litre $=$ 1.0 kg; s.h.c. $= 4\,200$ J/(kg K)]

(b) Assuming there is no change in the thermal energy store in the machine or air, how long does it take to heat the water?

30.7 An energy of 75 600 J must be supplied to raise the temperature of a 2.0 kg block of ice, initially at $-18\,°C$, to its melting point. Calculate the specific heat capacity of ice suggested by these figures.

30.8 Calculate the specific heat capacity of a 3.0 kg piece of metal which experiences a temperature rise of $25\,°C$ when it is heated at a rate of 60 W for 10 minutes, if a total of 3 000 J is lost to the surroundings during this process.

30.9 Assuming the system is well insulated, what temperature would 490 g of water reach, starting at 15 °C, if a 60 W heater heated it for 20 minutes? [Specific heat capacity of water = 4 200 J/(kg K)]

Additional Thermal Energy Questions

30.10 Calculate the thermal energy needed to raise the temperature of a 2.5 kg block of ice to its melting point, if it is taken from a freezer at −18 °C. [c_{ice} = 2.1 kJ/(kg K)]

30.11 The specific heat capacity of iron is 440 J/(kg K). How much thermal energy is gained when the temperature of 800 g of iron is raised by 120 °C?

30.12 Copper has a specific heat capacity of 390 J/(kg K). A 20 g piece of copper at 1 050 °C is dropped into a very large tank of water which is at 15 °C. What is the change in the thermal energy of the water when the copper has cooled to the temperature of the water?

30.13 When 30 g of gold is warmed by 50 °C, the change in its thermal energy store is 260 J. From these figures calculate the specific heat capacity of gold.

30.14 10 kg of water at 40 °C is mixed with 10 kg of water at 20 °C in a bath. Assuming that the total energy stored in the water does not change (none is lost to the surroundings), what will be the final temperature of the water? Hint: imagine energy leaving the "store" of the hotter water to increase the energy "stored" in the colder water, until the temperatures are equal.

30.15 A 1.50 kg lump of aluminium [c = 910 J/(kg °C)] at 100 °C is dropped into a beaker containing 1.00 kg of water at 20.0 °C. Assuming the system is well insulated, what temperature will the aluminium and water be when in thermal equilibrium?

30.16 Babies must be bathed at exactly 37 °C. A bath contains 10 kg of water at 15 °C. How much water at 50 °C needs to be added to make the temperature correct for bathing a baby? Hint: in your working, take the mass of hot water to be m, and write down equations which later enable you to work out the value of m.

31   Latent Heat

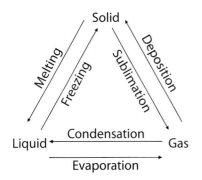

The three most commonly encountered states of matter are solids, liquids and gases.

When a substance changes state, it does not change temperature but thermal energy is still transferred.

The energy needed to change the state of a substance is called latent heat.

Specific latent heat of fusion, L, is the energy transferred from 1 kg of a substance changing from liquid to solid at a constant pressure. [unit: J/kg]

Specific latent heat of vaporisation is the energy transferred to 1 kg of a substance changing from liquid to gas at a constant pressure.

Equation:

thermal energy transferred for a change of state $=$ mass $\times$ specific latent heat

$$Q = mL$$

	Latent heat of fusion	Latent heat of vaporisation
Melting	Energy gained by substance	—
Freezing	Energy lost to surroundings	—
Evaporating	—	Energy gained by substance
Condensing	—	Energy lost to surroundings

> Example – The specific latent heat of fusion of ice is 3.36×10^5 J/kg.
> How much thermal energy is transferred to melt 2.00 kg of ice?
> $Q = mL = 2.00 \times 336\,000 = 672\,000$ J $= 672$ kJ

31.1 Work out the missing measurements from the following table.

Q	m	L
8.38 MJ	(a)	838 000 J/kg
251 kJ	0.75 kg	(b)
(c)	100 g	449 000 J/kg
740 mJ	10.0 mg	(d)
1.09 MJ	(e)	199 000 J/kg

31.2 A student measures 250 g of water and pours it into a beaker. They boil the water over a Bunsen burner for five minutes, then measure the mass of the water again; this time it is 200 g. The specific latent heat of vaporisation of water is 2 260 kJ/kg. How much energy has been transferred in evaporating the water?

31.3 Pure water boils at 100 °C, has a specific latent heat of vaporisation of 2 260 kJ/kg and a specific heat capacity of 4 200 J/(kg K).

(a) How much energy is required to boil away 2.0 kg of water if it is already at 100 °C?

(b) How much energy is required if the water started at 40 °C? [Hint: you need to use the heat capacity of water to work out the energy needed to raise the temperature of the water – see section 30 on page 91.]

31.4 1 000 kg of steam is condensed back to water in the condenser of a power station each hour. The specific latent heat of vaporisation of water is 2 260 kJ/kg. Calculate the energy output to the environment this causes each second.

31.5 A typical fluid used in a fridge has a latent heat of vaporisation of 200 kJ/kg. The fluid needs to remove 30 J from the fridge each second, and it does this by boiling alone. Calculate the minimum mass of fluid which must flow through the fridge each second.

31.6 How much energy would be required to enable 5.0×10^{-3} kg of ethanol to evaporate? The specific latent heat of vaporisation of ethanol is 840 kJ/kg.

31.7 A sample of solid ethanoic acid is at its melting point of 17.0 °C. It has a specific latent heat of fusion of 192 000 J/kg. How much ethanoic acid can be melted with 864 kJ of thermal energy?

31.8 Liquid nitrogen boils at −196 °C. 40.0 kg of liquid nitrogen in a dewar flask completely evaporates when 7.96 MJ of thermal energy is transferred. What is its latent heat of vaporisation?

31.9 A 1000 W heater is placed in an insulated beaker containing 750 g of water at 100 °C. The water vapour is allowed to escape. Assume that there is no loss to the surroundings via conduction, convection or radiation. The specific latent heat of vaporisation of water is 2 260 kJ/kg.

  (a) How much water is left after 5.0 minutes?

  (b) How long will it take for half of the water to have evaporated?

$^{12}/_{15}$

32   Payback Times ♡

Domestic photovoltaic solar panels or small scale wind turbines are popular additions to many people's homes. Given the ever rising cost of fossil fuels and their environmental impact, individuals are investing significant sums of money in the hope of saving money in the long run.

Payback time: the time to save as much money as the initial investment.

Example 1 – A wind turbine costs £1 000 including installation. Since its installation, it has saved the owner £50 per year on their electricity bill. How long will it take for the owner to be in profit?
£1 000/£50 per year = 20 years.

32.1 Two of the domestic upgrades with highest installation costs are solar panels and double glazed windows, whereas cavity wall insulation is more midrange.

(a) A photovoltaic solar panel costs £5 000 to install and saves £100 per year in electricity costs. What is the payback time?

(b) Fitting an entire house with double glazed windows costs £9 000. Once fitted, the windows provide a saving of £800 each year in the heating bill. What is the payback time?

(c) Fitting cavity wall insulation to a flat costs £400. The annual saving in heating bills is £90. What is the payback time?

32.2 A jacket for a hot water tank costs £30 and the payback time is 8.0 months. How much money does the jacket save each year?

32.3 Photovoltaic solar panels cost £500 per square metre to install on a 6.0 m × 6.0 m roof. They save the owner £75 per month. What is the payback time?

32.4 A builder quotes that they can fit loft insulation to your house for £300 and it will pay for itself in two years.

(a) How much money will you save in heating bills over ten years if their quote is accurate?

(b) What is the profit after ten years (money saved in heating minus the installation cost) if you follow their advice?

32.5 Another photovoltaic cell costs just £3 000 to install but saves only £50 per year in electricity costs. What is the payback time?

	Installation Cost	Annual Saving on Bill
A.	£10 000	£500
B.	£4 000	£250
C.	£500	£10

32.6 Three solar panels A, B, C have the above costs and savings. Calculate the payback time for each, and identify the best investment.

(a) Which is the best investment if you wish to pay for the solar panels as quickly as possible?

(b) Faced with this choice, which option would save you the most money over 30 years?

Example 2 – A wind turbine can generate an average of 100 W throughout the year [1 W = 1 J/s; see section 33]. Electricity suppliers charge for each kilowatt-hour (kW h):
1 kW h = 1 kW x 1 hour = 1000 W x 3600 s = 3.6×10^6 J
The owner usually pays 20p per kW h. If she wants to be "in profit" within 5.0 years, what is the maximum cost of the turbine?

$(100 \text{ W}/1\,000) \times 365$ days $\times 24$ hours per day $= 876$ kW h per year.
876 kW h $\times$ £0.20 $=$ £175.20 saved per year.
£175.20/year $\times 5.0$ years $=$ £876 maximum cost.

Assume one year $= 365$ days in the following questions.

32.7   A wind turbine can generate 150 W on average and costs £3 000.
   (a) What is the cost per kW h from the turbine by the end of the first year?
   (b) What does the cost per kW h fall to by the end of the fifth year?

   (c) If the lowest price for electricity from the utilities supplier is 15p per kW h, what is the payback time?

32.8   A photovoltaic phone charger, of voltage 5.00 V and current 2.00 A, can only operate between the hours of 8am and 7pm. It costs £40.15 and saves 10.0p per kW h. What is the payback time?

32.9   A wind turbine on a caravan can keep a 35 W fridge running. The payback time is 7.0 years and saves the owner 15p per kW h. How much did the turbine cost?

32.10 A nightlight has a power of 2.0 W. It only needs to be on from 7pm to 7am, but the parent does not want to switch it manually. A timer 'uses' electricity all the time. What is the maximum power of a timer which will still save the family money on their electricity bill?

33   Doing Work, Potential Energy and Power

Doing work always requires a force. However, applying a force does not necessarily mean that work is done.

A force does need an energy supply if the force's point of application is moving (unless the motion is at right angles to the force).
Example of a force which does require an energy supply:
Thrust from jet engine.
Example of a force which does not require an energy supply:
The weight of a book sitting on a table.
If the force is in the same direction as the motion, the force does work on the object and the object will speed up (unless other forces act).

If the force on the object is in the opposite direction to its motion, the object does work and it will slow down (unless other forces act).

When work is done, one energy store will decrease, and another will increase. Work is measured in joules (J) - the same unit as energy.

If the force is perpendicular to the motion, the object neither slows down nor speeds up. No work is done, and there is no energy transfer. However, the object will change direction and accelerate.

work done $=$ force $\times$ distance moved parallel to force

$$W = Fs$$

Energy Transfer, E $=$ Work Done, W

So, lifting a 1 N weight 1 m upwards requires work of 1 J.
Lifting a 1 N weight 2 m upwards requires work of 2 J.
Lifting a 2 N weight 2 m upwards requires work of 4 J.
Lifting a 10 N weight 4.0 m upwards onto a shelf, and then sliding it sideways by 2.0 m against a friction force of 2.5 N requires work of $40 + 5 = 45$ J.

The energy change each second is called the power, measured in watts (W).

power $=$ energy transfer$/$time

$$P = \frac{E}{t} \quad \text{or} \quad P = \frac{W}{t}$$

Example 1 – Calculate the power needed to push a car 12.5 m along a road with a force of 2 340 N in 15.0 s.

Work done $= Fs = 2\,340$ N $\times$ 12.5 m $= 29\,250$ J
Power $= W/t = 29\,250$ J$/15.0$ s $= 1\,950$ W

Example 2 – Calculate the energy transfer when a 20 kg sack of flour is winched 13.5 m upwards in a mill.

Force $=$ weight $=$ mass $\times g = 20.0$ kg $\times$ 10 N/kg $= 200$ N
Work done $= Fs = 200$ N $\times$ 13.5 m $= 2\,700$ J

Notice that the work done during lifting equals the increase in gravitational potential energy.

gravitational potential energy (GPE) $=$ mass $\times g \times$ height

$$E = mgh$$

33.1 Complete the table below, where each row is a separate question. The units are in the heading except in the cases where they are stated.

Force (N)	Distance (m)	Work (J)	Time (s)	Power (W)
4.5	8.2	(a)	10.0	(b)
(c)	30.0	12 000	(d)	7 200
650	75 cm	(e)	0.63	(f)
2 500	12.0	(g)	(h)	713
Weight for a mass of 100 kg	200 km	(i)	7.0 min	(j)

33.2 A builder needs to drag a sack of cement 20 m along the floor against a friction force of 60 N.

(a) Calculate the work done needed.

(b) The dragging took two minutes. What was power of the builder?

33.3 How much gravitational potential energy is lost when a 60 kg boy walks down a flight of stairs which is 4.5 m high?

33.4 A weight-lifter raises a barbell of mass 20 kg, doing 490 J of work on it. Through what height does she lift the barbell?

33.5 A lighting bar on stage has a mass of 300 kg when supporting stage lights.

(a) What is its weight?

(b) How much energy do you need to lift it by 10 m?

(c) If your power is 100 W, how long would it take you to lift the bar by 10 m?

(d) What is the increase in gravitational potential energy when the bar is lifted by 10 m?

33.6 A car park has three floors. The ground floor is at street level. The first floor is 4.0 m above the ground floor, the second floor is 3.0 m above the first floor.

(a) How much energy does it take to lift an 800 kg hatchback to the top level from the street?

(b) How much energy does it take to lift a 2000 kg SUV to the first floor from the street?

(c) Calculate the change in gravitational potential energy when a 400 kg city car moves to the top floor.

(d) ♡ Calculate the change in gravitational potential (that is, the gravitational potential energy per kilogram) in cases (a), (b) and (c) of this question.

33.7 When an object falls in a gravitational field it loses gravitational potential energy.

(a) How much gravitational potential energy is lost when a rock of mass 3.0 kg falls to the foot of a 250 m cliff on the Moon where 'g' is 1.6 N/kg?

(b) In a hydroelectric power station, how much potential energy is lost by 100 tonnes of water flowing down through the pipes, falling a vertical distance of 200 metres? (1 tonne $= 1\,000$ kg)

33.8 A rower exercises with an output power of 200 W. Her boat (containing 7 other equally-strong rowers) travels at a speed of 9.0 m/s.

(a) How far will the boat go in 20 s?

(b) How much energy will the crew have converted in 20 s?

(c) Calculate the force each rower exerts to push the boat along.

(d) What happens if you multiply the force in (c) by the speed of the boat?

Q33.8(d) should show you that there is another useful equation:

$$\text{power} = \text{force parallel to motion} \times \text{speed}$$

$$P = Fv$$

33.9 ♡ A car is being driven down a drag race track.

(a) Calculate the power of car engine needed if the car is to be driven at a constant 70 mph (31 m/s) against a combined friction and air resistance force of 1400 N.

(b) Calculate the power of car engine required to drive the car at a speed of 100 mph (44.7 m/s) against 1400 N of resistive force.

(c) In practice, the car needs a much more powerful engine. Why?

33.10 A mountain climber has a mass (with all of their equipment) of 95.0 kg. If they were perfectly efficient, and ate 500 g of chocolate (11.1 MJ of chemical energy), how high could they climb?

$^{24}/_{32}$

34 Kinetic Energy

The kinetic energy associated with a moving object depends upon its mass and speed.

Kinetic energy is a scalar quantity, which means that it has a magnitude but no direction.

Kinetic energy is measured in joules (J).

Numerically, if an object has 400 J of kinetic energy it will require a 400 N force to stop it in 1 m, as work done = force × distance ($W = Fd$).

Formula:

$$\text{kinetic energy} = \tfrac{1}{2} \times \text{mass} \times \text{speed}^2$$
$$E = \tfrac{1}{2} mv^2$$

(you can see where this formula comes from if you do Q34.9)

Suppose an object has 400 J of kinetic energy.

- The energy of an object with twice the mass, but the same speed, would be 800 J (double) because kinetic energy is proportional to mass.

- The energy of an object with twice the speed, but the same mass, would be 1600 J (quadruple) because kinetic energy is proportional to the square of the speed.

Example 1 – A 2.00 kg carton of milk is falling at 2.50 m/s. Calculate its kinetic energy.

$E = \tfrac{1}{2} mv^2 = \tfrac{1}{2} \times 2.00 \text{ kg} \times (2.50 \text{ m/s})^2 = 6.25 \text{ J}$

Example 2 – A 4.50 kg rolling skateboard has a kinetic energy of 32.0 J. How fast is it going?

$E = \tfrac{1}{2} mv^2$, so 32.0 J $= \tfrac{1}{2} \times 4.50 \text{ kg} \times v^2$

Therefore $32.0 = 2.25v^2$
$32/2.25 = v^2 = 14.2$, so $v = 3.77$ m/s

Example 3 – How much force will it take if you wish to stop a 930 kg car going at 14.5 m/s in a distance of 23.0 m?

$E = \frac{1}{2}mv^2 = \frac{1}{2} \times 930$ kg $\times (14.5$ m/s$)^2 = 97\,800$ J
Energy transferred $=$ force $\times$ distance, so $97\,800$ J $= F \times 23.0$ m,
$F = 97\,800$ J$/23.0$ m $= 4\,250$ N

34.1   Calculate the kinetic energy of a 2.0 kg motion trolley going at 3.0 m/s.

34.2   Calculate the kinetic energy of an 800 kg car when it is going at

(a) 30 mph (which is 13.4 m/s);

(b) 40 mph (which is 17.9 m/s).

(c) Road safety campaigners are continually reminding motorists that 40 mph is much more dangerous than 30 mph even though it only seems a little bit faster. What does this question suggest about the issue?

34.3   (a) Calculate the kinetic energy of a 20 tonne bus travelling at 40 mph. (1 tonne $= 1\,000$ kg)

(b) Calculate the kinetic energy of a 600 kg Formula 1 race car going at 83 m/s (about 190 mph), and compare it to that of the bus.

34.4   A 500 kg pumpkin is dropped 15 m on top of a school bus.

(a) How much gravitational potential energy was gained when it was winched up 15 m?

(b) Assuming all of this GPE is turned into kinetic energy as the pumpkin drops, work out its speed as it hits the school bus.

(c) Would the speed be any different for a 5.0 kg pumpkin?

34.5   At what speed is a 250 g stone moving if its kinetic energy is 3.5 joules?

34.6   What is the mass of an object travelling at 8.0 m/s which has 96 J of kinetic energy?

34.7   A car of mass 1200 kg slows down from a speed of 20 m/s to 10 m/s. How much kinetic energy does the car lose? [Hint: first work out the kinetic energy before and after the deceleration.]

34.8   How fast was a 1400 kg car travelling if it lost 280 kJ of kinetic energy in coming to a stop?

34.9   This question allows you to derive the equation for kinetic energy using a numeric example. We assume constant acceleration and no resistive forces.
You can use these equations:
distance = average speed × time
acceleration = change in speed / time taken
force = mass × acceleration
work done = force × distance

(a) A 700 kg car accelerates uniformly from rest to 30 m/s in 10 s. Calculate its acceleration.

(b) Calculate the force needed to give the car this acceleration.

(c) The average speed of the car is midway between the starting speed (0.0 m/s) and the final speed. Use this information to work out how far the car will go while accelerating.

(d) The kinetic energy equals the work done in accelerating it. Calculate the kinetic energy.

(e) Now repeat this question for a car of mass m going from rest to speed v in time t.

34.10 A 300 000 kg airliner is flying at 250 m/s at an altitude of 11 000 m. How large is its kinetic energy when expressed as a percentage of the total of the kinetic and gravitational potential energy?

35 Efficiency

To carry out an energy analysis of a physical event or process, we need to identify a clear start point and an end point. We then consider and calculate the changes in the energy stores at the start point and at the end point.

It is always true that there is no overall change in the total of all the energy stores - energy is conserved.

However, it is often the case that a process results in an overall increase in less-useful thermal stores (and a corresponding decrease in the total of the more-useful stores). What is meant by the word 'useful' depends on the situation. Sometimes it will be fairly obvious; sometimes you may be told; in some situations, you may have to think carefully.

For any given process (or system) we can calculate its efficiency. Efficiency has no units. It is usually written as a decimal (generally between 0.00 and 1.00), as a fraction or as a percentage.

efficiency $=$ useful energy transferred$/$total energy transferred

To express efficiency as a percentage, multiply the decimal answer by 100.

Sometimes the total energy transferred is the total electrical or mechanical work done.

Example 1 - An electric current drives an electric motor to raise a 25 N weight by a vertical distance of 1.2 m. The electrical work done by the power supply is 47 J. Calculate the efficiency of this process.

efficiency $=$ useful energy transferred$/$total energy transferred

$=$ GPE gained (or work done against gravity)$/$electrical work done

$$= 25 \times 1.2/47 = 0.64 \text{ (2sf)} \quad \text{or} \quad 64\%$$

> Example 2 - A battery powered motor is used to lift a load. As the load is lifted, the increase in gravitational potential energy is 230 J. The decrease in the energy stored chemically in the battery is 290 J. Calculate the efficiency of this process.
>
> efficiency = useful energy transferred/total energy transferred
>
> = increase in gravitational store/decrease in chemical store
>
> $= 230/290 = 0.79$ (2sf) or 79%

35.1 A battery powered toy car accelerates from 0.0 m/s to 8.5 m/s. Its mass is 0.55 kg. The chemical store in its battery is decreased by 30 J. Calculate the efficiency of this process.

35.2 A mains operated motor raises an 8 500 N weight to a height of 2.7 m. The electrical work done by the mains supply is 29 000 J. Calculate the efficiency of this process.

35.3 A mains powered electric winch pulls a trolley up a ramp, raising it by a vertical distance of 1.2 m. The trolley's weight is 6 200 N. The electrical work done by the mains supply is 9 350 J. Calculate the efficiency of this system.

35.4 Work out the missing measurements from the following table, where each row is a separate situation.

Efficiency	Energy in	Useful energy out	Wasted energy out
(a)	500 J	(b)	250 J
(c)	(d)	180 J	200 J
16.0 %	1.28 kJ	(e)	(f)
2.80 %	(g)	1.68 kJ	(h)

35.5 A student plugs her phone in for an hour to charge the battery. The power supply does 11 000 J of electrical work and the energy stored chemically in her phone battery increases by 8 300 J .

(a) Calculate the efficiency of this process.

(b) Calculate the increase in thermal energy resulting from 1.0 hour of charging her phone.

In the first minute of charging, it is reasonable to assume that all of the increase in the thermal energy raises the temperature of her phone battery. The battery has a mass of 28 g and its specific heat capacity is 480 J/(kg°C).

(c) Calculate the increase in the thermal energy in the first minute.

(d) Calculate the battery's temperature rise during the first minute.

Efficiency can also be calculated by considering power for a process. Then

$$\text{efficiency} = \text{useful power output}/\text{total power input}$$

To express the efficiency as a percentage, again multiply the decimal answer by 100.

Example 3 - An electric water heater heats water with an output power of 2 050 W whilst its electrical power input is 2 200 W.

$$\text{efficiency} = \text{useful power output}/\text{total power output}$$

$$= 2\,050/2\,200$$

$$= 0.93 \text{ (2sf)} \quad \text{or} \quad 93\%$$

35.6 A mains transformer has an input power of 2.0 kW and is 90% efficient. How much energy would be wasted in 10 minutes?

35.7 A machine has an efficiency of 60%, the useful power output is 150 W. What is the total input power?

35.8 An electric motor has a power input of 10 watts when lifting a weight with a pulley system. The motor and pulley system is 80% efficient. Calculate how much potential energy would be gained by the weight in 5.0 s.

35.9 An electric motor has a power input of 3.0 watts when lifting a weight. The weight gains 10 joules of potential energy in 5.0 seconds.

(a) What is the useful output power of the motor?

(b) What is the motor's efficiency in carrying out the operation?

35.10 A model hydroelectric power station produces just enough electric power to light a 6.0 W lamp. The model is 80% efficient at converting the potential energy store of the water into electrical work. What is the input power of the water running through the pipes?

Additional Efficiency Questions

35.11 An electric motor draws 2.0 A from a 12 V supply. It can lift a weight so that the weight gains 54 J of potential energy in 3.0 s. Calculate:

(a) the input power of the motor;

(b) the useful output power of the motor and;

(c) the efficiency of the motor.

35.12 A water pump, rated at 12 V; 5.0 A raises 30 kg of water through a height of 2.0 m in a time of 15 seconds. Calculate the pump's efficiency. [Assume the water has no kinetic energy on reaching the top.]

35.13 A hydroelectric power station generates 64 MW of electric power when the input power from the falling water is 70 MW. Calculate the efficiency of the system.

35.14 What is the efficiency when in standby mode of

(a) A modern television with a 'standby' electrical power of 0.30 W?

(b) An older television with a 'standby' electrical power of 5.0 W? Please note that the television in Q14.a is better than that in Q14.b, but efficiency percentages do not give you this information.

36   Power and the Human Body ♡

Formulae:

work done (J) $=$ force (N) $\times$ distance moved parallel to direction of force (m)

$$E = Fs$$

power (W) $=$ energy (J)/time (s)

$$P = \frac{E}{t}$$

36.1   A weight-lifter lifts a mass of 20 kg through 1.5 m ten times in one minute.

(a) What is the weight of the 20 kg mass?

(b) Calculate the total work done in lifting the weights.

(c) Calculate the weight-lifter's average power in watts.

36.2   A 70 kg bricklayer needs to put 100 bricks (2.0 kg each) on the first floor of some scaffolding. The first floor of the scaffolding is 3.0 m above the ground floor.

(a) Calculate the potential energy change when the bricks are lifted to the first floor.

(b) Assuming that they move five bricks at a time, calculate the energy needed for the bricklayer to climb the ladder enough times to lift 100 bricks to the first floor.

(c) If the bricklayer's maximum power is 800 W, how much time would they spend going up the ladder while doing the jobs in parts (a) and (b) combined?

36.3   The power needed to keep a human being alive is called the basal metabolic rate (BMR). For adults this is about 6.0 MJ/day.

(a) Calculate the BMR in watts (J/s).

(b) Calculate the BMR in joules per hour (J/h).

36.4 A cyclist on an exercise bike has a basal metabolic rate of 100 W. Her muscles are 30% efficient. This means that for every 100 J of energy given to the muscles in the form of food, only 30 J are converted into work done on the bike. The power reading on the exercise bike is 150 W. Calculate the total power needed by her body to produce this output.

36.5 Sometimes energies are given in kilocalories (kcal). One kilocalorie (sometimes called a Calorie) is equal to 4 200 J.

(a) Express the amount of energy needed to keep a person alive (6.0 MJ/day) in kcal/day.

(b) How much energy (in kcal) is needed per hour to stay alive?

36.6 Someone swimming has a total metabolic rate of 600 W.

(a) How much energy would they need for an hour of swimming? Give your answer in kilojoules (kJ).

(b) How much energy (in kcal) is needed for an hour of swimming?

36.7 The chemical processes in your body generate thermal energy, which keeps you warm. If you lose 30 J of thermal energy each second to your surroundings, your body needs to convert another 30 J into thermal energy each second to maintain body temperature. If this doesn't happen, your body temperature will fall, and you may become ill.
Fred's body has a surface area of 2.0 m². He loses 80 J of thermal energy each second to his surroundings.

(a) What basal metabolic rate is needed (in J/s) for Fred to keep his body temperature constant?

(b) How much thermal energy does Fred lose per second through each square metre of body surface?

(c) Fred's baby sister has a surface area of 0.20 m². How much thermal energy do you expect her to lose each second?

(d) Who finds it easier to stay warm - Fred or his baby sister? Why?

37 Springs and Elastic Deformation

extension = (length of a spring or material) − (its unstretched length)

For a spring or any material below its limit of proportionality, the force stretching (or compressing) the material is proportional to its extension (or compression). Twice the force causes twice the extension. This is Hooke's Law.

When the force is removed, it goes back to its original length. This is called elastic deformation.

The spring constant k measures the force needed to stretch it by 1 cm or 1 m. The unit of the spring constant is N/cm or N/m.

Formula:

$$\text{force (N)} = \text{spring constant (N/m)} \times \text{extension (m)} \qquad F = kx$$

For extension, two symbols are commonly used, e and x.

Example 1 – A spring is 5.0 cm long when unstretched. With a 6.0 N force stretching it, it becomes 13 cm long. What is the spring constant?

extension = 13.0 cm − 5.0 cm = 8.0 cm.

spring constant = force/extension = 6.0 N/8.0 cm = 0.75 N/cm
or 6.0 N/0.080 m = 75 N/m

Example 2 – What will the length of this spring be when it is stretched with a 9.0 N force?

extension = force/spring constant = 9.0 N/(0.75 N/cm) = 12 cm
length = 5.0 cm + 12 cm = 17 cm.

37.1 A spring takes a force of 5.0 N to extend it by 2.0 cm.

(a) What is the spring constant in N/m?

(b) What force is needed to extend it by 10 cm?

(c) What force is needed to extend it by 12 cm?

(d) What force is needed to extend it by 5.0 mm?

(e) What is the extension when the force is 45 N?

(f) What is the extension when the force is 7.0 N?

37.2 A spring is 6.0 cm long when it is not stretched, and 10 cm long when a 7.0 N force is applied.

(a) What is the spring constant in N/m?

(b) How long will it be with a 21 N force?

(c) How long will it be with a 5.0 N force?

(d) What force is needed to make it 20 cm long?

37.3 A company is making newton meters using 150 N/m springs.

(a) How far from the 0.0 N mark will the 10 N mark need to be?

(b) How far apart will the 4.0 N and 9.0 N marks be?

When two springs support a weight in series (one hanging off the other), they each carry the full weight of the load. When two identical springs support a weight in parallel, they share the weight of the load, but have the same extension as each other.

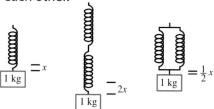

37.4 Two identical springs each have a spring constant of 200 N/m. A mass of 9.7 kg is hung from one spring, which is supported by the other. The top spring is supported by a strong beam.

(a) What is the total extension of the springs?

(b) The two springs are now set up in parallel to share the load.

What is the extension of each spring when the mass is supported?

Potential energy stored in a stretched spring

$$\text{work done when stretching a spring} = \text{force} \times \text{distance}$$

However, the force changes as you stretch the spring. To start with, very little force is needed to stretch it. At the end, the force is $F = kx$ where x is the final extension. The average force is $\frac{1}{2}kx$.

$$\text{work done in stretching a spring} = \text{average force} \times \text{distance}$$
$$= \tfrac{1}{2}kx \times x = \tfrac{1}{2}kx^2$$

$$\text{elastic potential energy (J)} = \tfrac{1}{2} \times \text{spring constant} \times \text{extension}^2$$
$$E = \tfrac{1}{2}kx^2$$

Example 3 – Calculate the elastic potential energy stored when a 1 000 N/m spring is stretched by 3.0 cm from its natural length.

With energy calculations, you should always use distances in metres.
Energy $= \tfrac{1}{2}kx^2 = \tfrac{1}{2} \times 1\,000\ \text{N/m} \times (0.030\ \text{m})^2 = 0.45\ \text{J}$

37.5 Calculate the elastic potential energy stored when a 1 000 N/m spring is stretched by 6.0 cm from its natural length.

37.6 Calculate the elastic potential energy stored when a 1 000 N/m spring is stretched by a 20 N force. [Hint: first work out the extension.]

37.7 Calculate the extension of a 500 N/m spring when storing 3.0 J.

37.8 *[Harder]* Calculate the elastic potential energy when a 40 N force stretches a spring by 0.50 cm.

37.9 How much strain energy is

(a) stored in a 40 kN/m spring when it is stretched by 3.0 cm?

(b) released on now relaxing to a stretch of 1.5 cm?

$^{14}/_{18}$

Waves and Optics

Waves can transfer energy (or information) without transferring material.

All waves involve oscillations (repeating motions back and forth).

Longitudinal Wave	Examples
Oscillations parallel to the direction of energy transfer	Sound Ultrasound Seismic P wave

Transverse Wave	Examples
Oscillations perpendicular to the direction of energy transfer	Light Other electromagnetic wave (e.g. radio) Seismic S wave

Wavelength	The distance from one peak to the next
Time period	The time for one whole wave to go past you
Amplitude	The height of the wave's peaks from the midpoint
Frequency	The number of waves going past each second
Peak	The highest point on the wave
Trough	The lowest point on the wave
Speed	How fast the wave goes

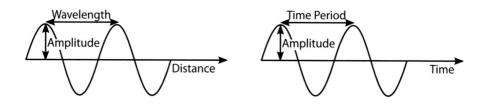

Formulae:

$$\text{wave speed} = \text{frequency} \times \text{wavelength} \qquad v = f\lambda$$

Reason: Length of wave made each second = number of waves made each second × length of each wave.

$$\text{frequency} = 1/\text{time Period} \qquad f = 1/T \qquad \text{so} \qquad T = 1/f$$

Reason: the frequency tells you how many time periods there are in one second, so multiplying the time period by the frequency will always give the answer 1.

38.1   Use the wave equations to fill in the blanks in the tables:

Time period (s)	Frequency (Hz)	Frequency (Hz)	Speed (m/s)	Wavelength (m)
0.10	(a)	(f)	300	30
0.050	(b)	(g)	300	15
0.0050	(c)	(h)	300	1.50
2.5	(d)	0.40	(i)	420
(e)	25	20	2.0	(j)

38.2   A musical note has a frequency of 440 Hz. The speed of sound in air is 330 m/s.

(a) What is the wavelength of the sound?

(b) What is the time period of the sound?

38.3   An ultrasound pulse has a wavelength of 1.0 mm. Its speed in water is 1 400 m/s.

(a) What is the frequency?

(b) What is the time period of the sound?

38.4   A lighthouse flashes once every 7.1 s. What is its frequency?

38.5   The mains power has a frequency of 50 Hz. What is its time period?

38.6 When a musical note goes up one octave in pitch, its frequency doubles. What happens to its wavelength?

38.7 What is the period of a wave whose frequency is 4.0 Hz?

When a wave moves from one material to another, the frequency does not change. If the speed changes, the wavelength will change too.

38.8 Green light has a wavelength of 0.000 000 50 m, and a speed of 300 000 000 m/s.

(a) Calculate its frequency.

(b) When the green light goes into glass, it slows down. Its new speed is 200 000 000 m/s. What is its frequency and wavelength in glass?

38.9 Calculate the frequency of a water wave which has a wavelength of 1.5 m and travels a distance of 10 metres in 5.0 seconds. [Hint: This question has two stages. You will have to work something out before you can calculate the quantity requested.]

38.10 A water wave, travelling at 2.5 m/s, has a wavelength of 50 cm. What is the period of the wave? [Hint: This question has two stages. You will have to work something out before you can calculate the quantity requested.]

Additional Wave Properties and Basic Equations

38.11 On a stormy day, a girl counts the number of wave crests breaking on the shore and finds that there are 60 in 4.0 minutes. Calculate the water waves' frequency in hertz (waves per second).

38.12 A tuning fork makes the musical note one octave above 'middle C' (in the 'scientific designation'). What is its frequency, if its prongs vibrate 2 560 times in 5.0 s?

38.13 What is the wavelength of a 200 Hz sound wave in the air if the speed of sound in air is 340 m/s?

38.14 What is the speed of sound through an aluminium rod if a sound vibration of frequency 13 kHz has a wavelength of 40 cm?

38.15 What is the frequency of a wave of red light in the air where its wavelength is 6.8×10^{-7} m?

38.16 Calculate the wavelength of Radio 4 which broadcasts on a frequency of 198 kHz.

38.17 Certain X-rays have a frequency of 1.0×10^{19} Hz. Calculate their wavelength in the air.

38.18 What is the wavelength of a radio station which sends out radio waves of frequency 1.15 MHz?

38.19 A certain radio station broadcasts on a frequency of 101.7 MHz. Calculate the wavelength of the radio wave.

38.20 Calculate the frequency, in kilohertz, of a radio station which broadcasts on the Medium Wave with a wavelength of 1 500 m.

39   Reflection – Plane Mirrors ♡

Reflections can be of two types: diffuse and specular. Diffuse reflections are from rough surfaces, where the light rays are scattered in all directions. Specular reflections are from smooth surfaces, where the law of reflection can be easily verified.

The law of reflection states that the angle of incidence is equal to the angle of reflection. The angle of incidence is the angle between the incident ray and the normal. The angle of reflection is the angle between the reflected ray and the normal at the point where the reflection occurs.

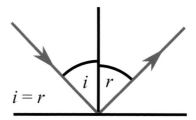

A normal is an imaginary line that is $90°$ to the surface.

An image is a point in space from where light rays can be considered to cross each other. Plane mirrors produce a virtual image - that is, an image that does not have light rays actually passing through it; the light rays appear to meet but they actually do not.

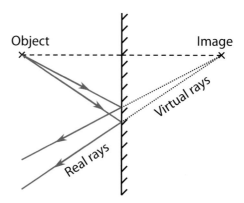

Virtual rays are extrapolated real rays. Light does not actually emerge from a virtual image, but an observer does not know that just by looking.

39.1 Copy and complete the following ray diagrams, each showing a ray incident on a plane mirror, by drawing in the reflected ray at the correct angle to the normal.

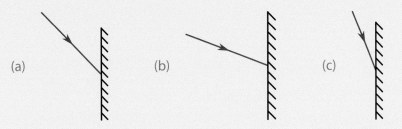

(a)   (b)   (c)

39.2 For each of the ray diagrams which show three rays of light incident on a plane mirror, copy and complete the diagram by drawing in the reflected rays.

Parallel Rays   Converging rays   Diverging rays

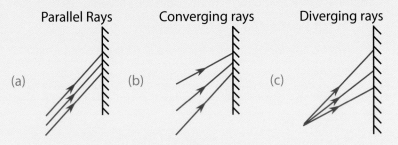

(a)   (b)   (c)

39.3 Copy and complete the ray diagram to show how the plane mirror makes a virtual image of the object (black arrow).

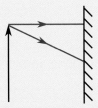

What do you notice about the distance of the image from the mirror compared to the distance of the object from the mirror?

39.4 An object is 10 cm away from a plane mirror. An observer is the same distance away from the mirror, and 10 cm away from the object.

(a) Sketch a ray diagram of the situation.

(b) How far has the light travelled from the object to the observer via the plane mirror?

(c) What is the angle of incidence?

(d) How far away from the image is the observer?

39.5 An object is placed at a mystery distance from a plane mirror. An observer is 15 cm away from the mirror. The angle of incidence is 15°. The ray that travels from the object to the observer is incident on the plane mirror a perpendicular distance of 4.0 cm from the imaginary line that connects the object to the image.

(a) Sketch a ray diagram of the situation.

(b) What is the length of the ray from the object to the mirror?

(c) What is the distance between the object and the image?

(d) What is the total distance the ray has travelled from the object to the observer?

(e) How far away from the observer is the image?

39.6 (a) What does the term 'lateral inversion' mean?

(b) Return to question 39.4, but now consider an object that is not just a point, but also has a shape. Answer part (a) of that question again; consider light from the parts of the object. How do these lead to a difference in appearance of the image compared with the object?

39.7 A common question is, 'why is writing, when viewed in a mirror, reflected from left to right, not from top to bottom?' What is the best response to this question, using the physics detailed here?

$^{15}/_{20}$

40 Reflection – Concave Mirrors ♡

The most commonly encountered curved mirrors are spherical. Spherical mirrors are either concave or convex.

Concave mirrors produce a real image at the focal point (labelled F in the diagram below) when parallel rays are incident parallel to the principal axis, which passes through the centre of curvature of the mirror (labelled C). The distance from the mirror to C is always double the distance from the mirror to F for spherical mirrors. In the diagram, arrows have not been included because the rays would follow the same geometric path in the reverse direction. An object at F will produce an image at infinity.

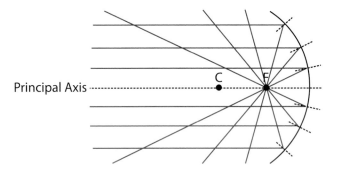

An object placed at C produces a real image at C because the rays are always incident on the mirror with zero angle of incidence. Combining these two ideas, the image of an object placed anywhere between C and F can be found graphically thus:

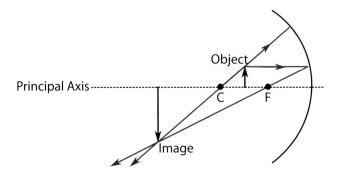

The three rules:
1. Rays passing through C reflect back through C.
2. Rays parallel to the principal axis reflect through the focal point.
3. Rays passing through the focal point reflect parallel to the principal axis.

For each of the questions, sketch the ray diagram.

40.1 Copy and complete the following ray diagrams, each showing rays incident on concave mirrors, by drawing in the reflected rays at the correct angles.

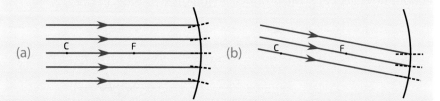

(a)   (b)

40.2 What is the name for the point to which rays parallel to the principal axis converge after being reflected from a concave mirror?

40.3 The diagram below shows rays approaching a concave mirror, having been generated at the mirror's focal point. Draw a diagram to show what happens to the rays after reflection from the mirror.

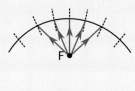

40.4 An object placed at the centre of curvature produces an image.

(a) What type of image is produced? [Real or virtual]

(b) Where is the image produced? [Between the mirror and F, at F, between F and C, at C, or between C and infinity]

(c) What is the orientation of the image? [Upright or inverted]

(d) What is the size of the image? [Diminished, same size as the object, or enlarged]

40.5 An object between C and F produces an image.
    (a) What type of image is it?　　(c) What is its orientation?
    (b) Where is the image?　　　　(d) What is its size?

40.6 An object placed between C and infinity produces an image.
    (a) What type of image is it?　　(c) What is its orientation?
    (b) Where is the image?　　　　(d) What is its size?

40.7 An object placed between F and the mirror produces an image.
    (a) What type of image is it?　　(c) What is its orientation?
    (b) Where is the image?　　　　(d) What is its size?

$^{15}/_{20}$

41   Reflection – Convex Mirrors ♡

The image formed from a convex mirror is always upright, diminished and virtual, regardless of where the object is placed.

Rays that are parallel to the principal axis reflect in a direction directly away from the focal point, which is halfway between the mirror and the centre of curvature for the mirror. When drawing ray diagrams, virtual rays can be drawn to correctly determine the path of the reflected rays.

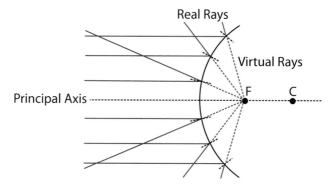

As with concave mirrors, a ray incident on the mirror with an angle of incidence of zero (through the centre of curvature) will reflect in the opposite direction with zero angle of reflection. The virtual ray extrapolated from the incident ray will pass through the centre of curvature.

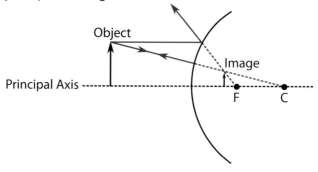

The two rules:
1. Rays which are incident in the direction of C reflect away from C.
2. Rays parallel to the principal axis reflect away from the virtual focal point.

For each of the questions, sketch a ray diagram.

41.1 Copy and complete the following ray diagram rays incident on a convex mirror, by drawing in the reflected rays at the correct angles.

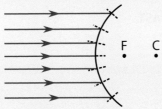

41.2 An object 3.0 cm long is placed 10.0 cm from a convex mirror. The radius of curvature of the mirror is 14.0 cm. Use a scale diagram to measure:

(a) The size of the image;

(b) Its distance from the mirror.

41.3 An object 1.5 cm long is placed 5.0 cm from a convex mirror. The radius of curvature of the mirror is 10.0 cm. Use a scale diagram to measure:

(a) The size of the image;

(b) Its distance from the mirror.

41.4 Identify the similarities and differences between concave and convex mirrors.

$^6/_8$

42   Refraction ♡

Light bends as it enters a glass block because the light travels slower in glass. This causes the wavelength of the light to get smaller, and causes the direction of the light to change.

We say light bends 'towards the normal' when it slows down, and bends 'away from the normal' when it speeds up.

Remember 'Light goes **FAST**.'

When it goes **F**aster
it bends **A**way from the normal

When it goes **S**lower
it bends **T**owards the normal.

Formulae for refraction are explained in Refractive Index & Snell's Law ♡ - P138.

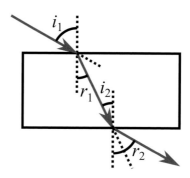

The direction can be correctly predicted by viewing the incoming light as a car whose wheels travel more slowly once they've crossed the boundary. If the front right wheel hits the boundary and slows down first, the car will turn right until the front left wheel also reaches the boundary.

refractive index $=$ speed of light in air / speed of light in material

The refractive index is always greater than or equal to 1.

42.1 The table shows some data for five materials. Calculate the refractive index for each one.

Material	Speed of light (m/s)	Refractive Index
Air	3.0×10^8	(a)
Glass	1.9×10^8	(b)
Water	2.3×10^8	(c)
Diamond	1.2×10^8	(d)
Turpentine	2.0×10^8	(e)

42.2 Does light *never, not usually, usually or always* bend towards the normal when going into a material with a higher refractive index with $i \neq 0$?

42.3 Would light with $i = 20°$ bend towards (T) or away from (A) the normal when passing from:

(a) air into water;

(b) water into glass;

(c) water into turpentine;

(d) glass into turpentine;

(e) diamond into air;

(f) turpentine into glass;

(g) turpentine into diamond;

(h) glass into water.

42.4 Violet light is slower in glass than red light. All colours of light travel at the same speed in air. A narrow, white beam of light enters a glass block with $i = 30°$. Which colour bends the most on refracting as it enters the glass block?

42.5 Different colours of light have different refractive indices in glass.

(a) Which has the lower refractive index - violet or red light?

(b) A wide beam of white light shines at an angle on a rectangular glass block, refracting on entry and on exit. Will the beam be parallel or diverging on leaving the block?

(c) The wide beam of white light now shines on a glass block that is triangular when viewed from above. Will the beam be parallel or diverging on leaving the block?

43   Total Internal Reflection ♡

The diagrams below show rays of light entering a glass block at different angles. The last one shows light hitting the boundary at a very glancing angle.

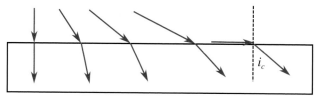

The next diagram shows the situation where light is leaving a glass block. Notice that these are identical to the rays shown above but with the direction reversed.

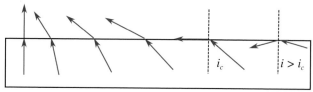

Where the angle of incidence is greater than i_c (the critical angle), the light cannot refract, and so it all reflects back inside the material.

Total internal reflection occurs when light attempts to leave a glass or Perspex block with an angle of incidence bigger than the critical angle. None of the light refracts. None of it leaves.

The critical angle for light leaving a glass block into air is $42°$.
The critical angle for light leaving water into air is $49°$.
The critical angle for light leaving diamond into air is $24°$.
The critical angle for light leaving cubic zirconia into air is $28°$.

The slower light travels in a material, the higher its refractive index, and the smaller its critical angle at an air boundary.

43.1   Given the previous critical angle data, put the five materials (air, glass, water, diamond, and cubic zirconia) in order of increasing speed of light.

43.2 The critical angle for the glass/water boundary is 62°. For TIR, from which medium must light hit the boundary?

43.3 Complete the table, stating whether **total internal reflection** (TIR) or **refraction** (R) will occur:

Light moving from	Light moving to	Angle of incidence	What happens?
Air	Glass	30°	(a)
Glass	Air	30°	(b)
Air	Glass	49°	(c)
Glass	Air	49°	(d)
Water	Air	43°	(e)
Glass	Water	70°	(f)
Water	Glass	82°	(g)
Diamond	Air	24°	(h)
Cubic Zirc.	Air	28°	(i)

43.4 The diagram shows a side view of a swimming pool. On the bottom are two people, P_1 and P_2, trying to see objects O_1, O_2 and O_3 outside the pool. The observers are also trying to see each other without looking directly at each other. Which of the rays shown are possibilities for observing?

(a) Is ray (a) possible?
(b) Is ray (b) possible?
(c) Is ray (c) possible?
(d) Is ray (d) possible?
(e) Is ray (e) possible?
(f) Is ray (f) possible?

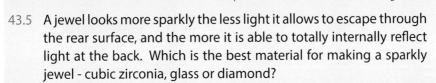

43.5 A jewel looks more sparkly the less light it allows to escape through the rear surface, and the more it is able to totally internally reflect light at the back. Which is the best material for making a sparkly jewel - cubic zirconia, glass or diamond?

$^{17}/_{22}$

44   Diffraction ♡

When waves encounter an obstacle or aperture (a gap), they spread out.

For gaps, the amount the waves spread out depends on the wavelength divided by the width of the aperture. In the two images below, waves are travelling from the top of the image to the bottom. The lines show the places where waves are currently at a crest, and are known as wavefronts. The wavelength is the same in both images (the distance between the waves fronts is the same), but the width of the gap is different. Notice that the diffraction angle, marked with the coloured lines, is greater for the narrower gap.

When waves are incident on an obstacle that is smaller than the wavelength of the wave, the waves diffract around the obstacle so very little shadow can be seen.

When waves are incident on an obstacle that is larger than the wavelength of the wave, the waves diffract around the edges of the obstacle but some of the wave energy reflects back from the obstacle and there is a shadow behind the obstacle.

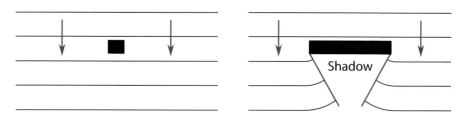

44.1 In the following table, label whether diffraction can be seen (Y) or not (N).

		Aperture Size		
		10.0 m	10.0 cm	0.01 mm
	10.0 m	-	(a)	(b)
Wavelength	10.0 cm	(c)	-	(d)
	0.01 mm	(e)	(f)	-

44.2 In the following table, label whether there is an obvious shadow behind the obstacle (Y) or not (N).

		Obstacle Size		
		5.0 m	25.0 cm	1.0 cm
	5.0 m	-	(a)	(b)
Wavelength	25.0 cm	(c)	-	(d)
	1.0 cm	(e)	(f)	-

44.3 Put these cases in order of size of diffraction angle, from largest diffraction angle to smallest, for a wave passing through an aperture. [Hint: the larger the wavelength divided by the aperture width, the greater the diffraction angle.]

Case	Wavelength	Aperture Width
1	550 nm	0.010 0 mm
2	700 nm	0.100 mm
3	1 400 nm	100 µm
4	5.00 cm	10.0 cm
5	15.0 cm	1 000 µm

44.4 A young astronomer has a telescope with a 6.0 cm diameter lens, and uses it to take pictures using visible light (wavelength = 500 nm). If a professional astronomer wanted images just as precise using 30 cm radio waves, what diameter of dish would be needed? (The main factor causing blurring in a good telescope is diffraction.) [Hint: use ratio and proportions.]

44.5 Copy and complete the following diagrams by drawing the position of the water wave fronts after they have been diffracted by passing through the gaps in the barriers.

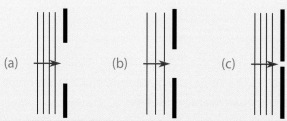

44.6 Copy and complete the following diagrams by drawing the position of the water wave fronts after they have been diffracted by passing the obstacles.

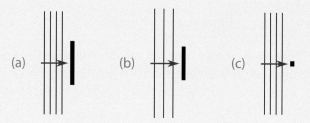

44.7 A person with perfect eyesight can only read a message written using 1.0 mm pixels if it is closer than 14 m (with a pupil diameter of 7.0 mm) because diffraction caused by the pupil blurs light from one pixel into another at greater distances. How far away could they read the same message if they used binoculars with 25 mm diameter lens? [Hint: use ratios or proportions.]

44.8 For satellite communications, radio and microwave transmission dishes need to be wide enough to prevent excess diffraction. Rank the following situations with the most parallel beam first.

(a) 3.0 cm microwave from a 21 cm radius dish.

(b) 15 cm radar from a 3.0 m radius dish.

(c) 500 nm light from a 8.0 μm blood cell.

45   Seismic Waves and Earthquakes ♡

When a geological plate moves suddenly during an earthquake, it sets off waves which travel through the rocks. Some waves travel along the Earth's surface (e.g. Love waves). Others, P and S waves, travel through the Earth.

Primary (P) Waves	Secondary (S) Waves
Longitudinal wave - oscillations parallel to direction of energy transfer.	Transverse wave - oscillations perpendicular to direction of energy transfer.
Faster (typical speed near surface of 8 km/s). These are the first waves to reach seismometers - this gives them their name 'primary'.	Slower (typical speed near surface of 5 km/s). These waves reach the seismometers later - this gives them their name 'secondary'.
Can travel through liquids (such as the Earth's outer core).	Cannot travel through liquids (such as the Earth's outer core) - this is because they are transverse.
Can be reflected at any boundary between two different regions.	
Can be refracted (or bent) by any change in rock compressibility or density. Generally as waves get deeper (and the pressure rises), their speed rises and they bend away from the 'normal' (the vertical). When P waves pass from the (solid) mantle to the (liquid) outer core, they slow down, and bend towards the vertical. When they reach the (solid) inner core, they speed up, and bend away from the vertical.	

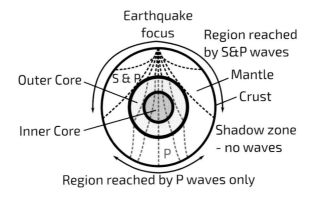

Region reached by P waves only

45.1 The average speed of S waves in the mantle is 6.0 km/s and the average speed of P waves in the mantle is 11 km/s. Ignore the crust, and treat the mantle as being 2900 km deep.

(a) How much time would an S wave take to travel from a seismic event and return to the focus after having reflected once from the mantle-outer core boundary?

(b) How much earlier would the reflected P waves be detected than the S waves?

Typical speeds of seismic waves and rock densities are shown in the table.

Region	Depth (km)	Density (kg/m³)	Speed (km/s)	
			P	S
Crust	$0 \sim 10$	3.0×10^3	8.0	5.0
Mantle	$\sim 10 - 2900$	$(3.0 - 5.0) \times 10^3$	$8.0 - 13$	$5.0 - 8.0$
Outer Core	$2900 - 5200$	10^4	$8.0 - 10$	-
Inner Core	$5200 - 6400$	1.2×10^4	11	$3.0 - 4.0$

(eqseis.geosc.psu.edu/~cammon/HTML/Classes/IntroQuakes/Notes/waves_and_interior.html)

Example 1 – What is the delay between receiving P and S waves travelling through the crust to a seismometer 200 km from the earthquake's focus?
Time for P wave = Distance / Speed = 200 km/8.0 km/s = 25 s.
Time for S wave = Distance / Speed = 200 km/5.0 km/s = 40 s.
Delay = 40 s – 25 s = 15 s.

Example 2 – If the delay between receiving P and S waves is 5 s, how far away is the earthquake's focus?
We call the distance d, taken in km where time will be in seconds.
For the P wave, the time take to arrive is $t_p = d/8$.
For the S wave, the time taken to arrive is $t_s = d/5$.
We are told the delay is 5 s. Thus $t_s - t_p = 5$.
Therefore $d/5 - d/8 = 5$. So, $0.2d - 0.125d = 5$, and accordingly $0.075d = 5$.
We finally get $d = 5/0.075 = 67$ km (70 km to 1sf).

45.2 For speeds of 5.0 km/s for S and 8.0 km/s for P waves, complete the table.

Distance	P wave Time	S wave Time	Delay
30 km	(a)	(b)	(c)
60 km	(d)	(e)	(f)
150 km	(g)	(h)	(i)
200 km	(j)	(k)	(l)
(m)	(n)	(o)	8.0 s

45.3 Explain how the location of an earthquake's focus can be worked out from distance measurements made from three seismometers.

45.4 For each of the following questions, give your answer as 'S wave', 'P wave', 'neither' or 'both'. Which seismic waves:

(a) are longitudinal?

(b) can travel through a part of the Earth which is liquid?

(c) can travel through the mantle?

(d) are generated by an Earthquake?

(e) will cause the rock next to the focus (at the same depth) to move up and down?

(f) will cause the ground above the focus to move up and down?

(g) will be detected by a seismometer in the 'shadow zone' of an earthquake?

(h) travel faster at a given depth in the mantle?

(i) arrive at a seismometer last from a particular earthquake?

(j) can change direction at a boundary between two parts of the mantle?

(k) can be detected on the opposite side of the Earth from the focus?

46   Refractive Index & Snell's Law ♡

Data:

$$\text{refractive index of glass} = 1.50$$
$$\text{refractive index of water} = 1.34$$
$$\text{refractive index of diamond} = 2.42$$
$$\text{refractive index of cubic zirconia} = 2.16$$
$$\text{refractive index of air} = 1.00$$
$$\text{speed of light in a vacuum} = 3.00 \times 10^8 \text{ m/s}$$

Refraction describes the change of direction of light on entering or leaving a material when it crosses the boundary.

Refraction is caused by the difference in the speed of light in the materials. To compare the speed of light in different materials, we compare their refractive indices.

$$\text{refractive index} = \frac{\text{speed of light in a vacuum}}{\text{speed of light in a material}} \qquad n = \frac{c}{v}$$

Air has a refractive index of 1.00, so the speed of light in air is very similar to the speed of light in a vacuum.

The larger the refractive index, the slower light travels.

Example 1 – Calculate the speed of light in diamond.

$$n = \frac{c}{v} \quad \text{so} \quad v = \frac{c}{n} = \frac{3 \times 10^8}{2.42} = 1.24 \times 10^8 \text{ m/s}$$

46.1   Calculate the speed of light in the following materials.

(a) What is the speed of light in glass.

(b) What is the speed of light in water.

46.2   The speed of light in hydrogen disulphide is 1.59×10^8 m/s. Calculate the refractive index of hydrogen disulphide.

Snell's Law enables us to calculate the angles when light refracts.

For light entering a material from air

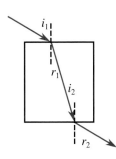

$$\sin(r_1) = \frac{\sin(i_1)}{n} \quad \text{so} \quad r_1 = \sin^{-1}\left(\frac{\sin(i_1)}{n}\right)$$

For light leaving a material to pass into air

$$n \times \sin(i_2) = \sin(r_2) \quad \text{so} \quad r_2 = \sin^{-1}(n \times \sin(i_2))$$

These formulae are explained on P141.

Example 2 – Light enters glass with an incident angle of 25°. Calculate the angle of refraction.

$$r = \sin^{-1}\left(\frac{\sin(i)}{n}\right) = \sin^{-1}\left(\frac{\sin(25°)}{1.50}\right) = \sin^{-1}(0.28) = 16°$$

46.3 Calculate r for light entering glass from air if:

(a) $i = 20°$;

(b) $i = 30°$;

(c) $i = 40°$.

46.4 Calculate r for light leaving glass into air if:

(a) $i = 20°$;

(b) $i = 30°$;

(c) $i = 40°$.

46.5 Light is incident on a boundary between water and air.

(a) Calculate r for light entering water if $i = 60°$.

(b) Calculate r for light leaving water if $i = 35°$.

46.6 In an experiment shining light through air into transparent waffle-cheese, $i = 60°$ and $r = 16°$.

(a) Calculate the refractive index (n) for wafflecheese.

(b) Calculate the speed of light in wafflecheese.

46.7 A ray of light makes an angle of incidence of $40°$ with the normal between air and a liquid. The angle of refraction in the liquid is $28°$. Calculate the value of the refractive index of the liquid.

46.8 The index of refraction of a kind of glass for a certain wavelength of red light is 1.51. It is 1.55 for violet. A ray of white light is incident on a prism made of the glass at $30°$ to the normal. Calculate the angle between the red and violet rays in the glass.

46.9 A ray of monochromatic light, travelling through air, makes an angle of $30°$ with the surface of a rectangular block of a certain type of transparent plastic. It makes an angle of $53°$ with the surface inside the plastic. What is the value of the plastic's refractive index?

46.10 [Harder] A block of glass is sitting on the bottom of a tank full of water. Light enters the glass from the water with an incident angle $i = 40°$. What will the angle of refraction (r) in the glass be? [Hint: if you like, you can pretend that there is a very small air gap between the glass and the water.]

When light passes from one material (with refractive index n_1) to a second material (refractive index n_2), then the general formula is

$$n_2 \sin(r) = n_1 \sin(i)$$

Notice that if the first material is air, $n_1 = 1$, then $n_2 \sin(r) = \sin(i)$.

If the second material is air, $n_2 = 1$, then $\sin(r) = n_1 \sin(i)$.

These agree with our earlier formulae, as well as the answer for Q46.10.

Reasoning behind Snell's Law

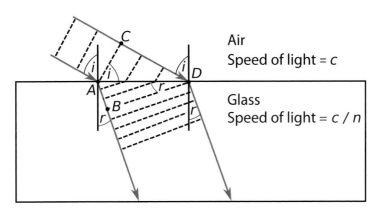

Wavefronts show the locations of crests of the wave at a particular time. The rays shown are the direction the waves travel and are perpendicular to the wave fronts. Here, the wavefronts are shown by dotted lines. The perpendicular distance between two wavefronts is the wavelength.

In the time (t) that lights travels from C to D, light also travels from A to B.

$$t = \frac{AB}{(c/n)} = \frac{CD}{c}$$

So $CD = n \times AB$ and $CD/AB = n$.

$\angle CAD = i$ and $CD = AD\sin(\angle CAD)$ so $CD = AD\sin(i)$

$\angle ADB = r$ and $AB = AD\sin(\angle ADB)$ so $AB = AD\sin(r)$

Dividing these equations gives $\dfrac{CD}{AB} = \dfrac{\sin(i)}{\sin(r)}$ but $\dfrac{CD}{AB} = n$.

Therefore $n = \dfrac{\sin(i)}{\sin(r)}$, so $\sin(r) = \dfrac{\sin(i)}{n}$.

This is Snell's Law.

47   Calculating Critical Angles ♡

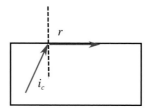

The conditions for total internal reflection are that the light

- must be attempting to leave a material into air, or more generally

    – crossing from a higher to lower refractive index material
    – this means that the light crosses a boundary where it speeds up

- and the angle of incidence must be above the critical angle (i_c).

If the angle of incidence were exactly critical, then the angle of refraction would be a right angle.

So $i = i_c$ and $r = 90°$. Remember, $\sin(90°) = 1$
Snell's Law for light leaving a material into air is

$$\sin(r) = n \sin(i)$$

In this case $\sin(90°) = n \sin(i_c)$. So $\sin(i_c) = \dfrac{1}{n}$ and $i_c = \sin^{-1}\left(\dfrac{1}{n}\right)$.

The refractive index $n = \dfrac{1}{\sin(i_c)}$.

Data:

refractive index of glass $= 1.50$
refractive index of water $= 1.34$

47.1 Calculate the critical angle for light leaving glass into air.

47.2 Calculate the critical angle for light leaving water into air.

47.3 The critical angle for light leaving diamond into air is $24°$. Calculate the refractive index (n) for diamond.

Where light goes from one material (refractive index n_1) to another (n_2), we use the more general form of Snell's Law.

$$n_1 \sin(i) = n_2 \sin(r)$$
$$n_1 \sin(i_c) = n_2 \sin(90°) = n_2$$
$$\Rightarrow \sin(i_c) = \frac{n_2}{n_1}$$

47.4 The critical angle for light leaving a particular type of glass is $38.4°$. What is its refractive index?

47.5 What is the critical angle for light passing from glass to water?

47.6 The inner section (core) of an optic fibre has a refractive index of 1.52 and the critical angle for light leaving the core into the cladding is $43.7°$. What is the refractive index of the outer section (cladding) that is keeping the light inside the core?

47.7 In entering a transparent material from the air, the wavelength of a laser's light decreases from 600 nm to 451 nm. Calculate the refractive index of the material. [*Hint: when a wave passes from one material to another, the frequency doesn't change. The effect of the speed change on the wavelength can be worked out from speed $=$ frequency $\times$ wavelength*]

47.8 A tube of glass of refractive index 1.65 is surrounded by glass of refractive index 1.51. Calculate the critical angle for light travelling along the tube and incident on the boundary between the glasses.

47.9 A thin ray of monochromatic light enters a block of pure ice at an angle of $42.0°$ to the normal from the air and the refracted angle in the ice is $30.7°$. Calculate the critical angle for ice.

48   Convex Lenses ♡

In the diagram below, the object has size O, the image size I, and the convex lens has a focal length f. We can work out the location of the image by drawing two rays through the system.

1. A ray passing through the centre of the lens is not bent.

2. A ray travelling parallel to the axis will bend at the lens, so that it crosses the axis at the focal point F (distance f behind the lens).

3. The image I is where the rays meet. If the rays are diverging (spreading apart) after the lens, extend both back to the left to find a place where the lines meet – this will be a virtual image.

4. The object distance is labelled u. The image distance is labelled v.

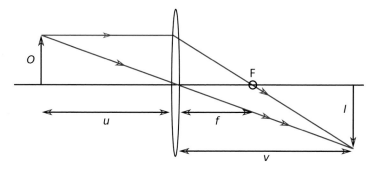

48.1   Copy and complete the ray diagrams by drawing the path of each ray of light after passing through the lens. Hence find the position, size and orientation of the image formed. Is it real or virtual?

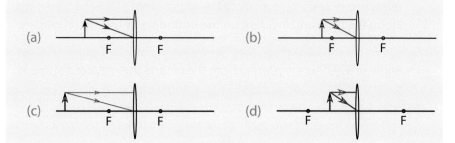

Power: The "strength" with which a lens focuses a parallel beam to a point (a focus) is measured as its power. Lens power is measured in dioptres (D).

$$\text{power in dioptres (D)} = (\text{focal length in metres})^{-1}$$

$$P = \frac{1}{f} = f^{-1}$$

Example 1 – Calculate the power of a lens with a 10 cm focal length.

$$10\,\text{cm} = 0.1\,\text{m} \quad P = 1/f = 1/0.1 = 10\,\text{D}$$

48.2 Use the equation $P = f^{-1}$ to work out the power in dioptres (D) of the following lenses.

   (a) A convex lens with a 1.6 m focal length.

   (b) A convex lens with a 50 cm focal length.

   (c) A convex lens with a 5.0 cm focal length.

48.3 Calculate the focal lengths in metres (m) for the following lenses.

   (a) A +2.5 D lens.

   (b) A +1.5 D lens.

   (c) A +20 D lens.

48.4 Calculate the focal lengths, in centimetres, of these converging lenses.

(a) +20 D	(c) +10 D	(e) +2.0 D
(b) +2.5 D	(d) +14 D	(f) +5.0 D

48.5 Calculate the power, in dioptres, of the following lenses.

   (a) A converging lens with a focal length of 40 cm.

   (b) A lens which brings parallel rays of light to a focus at a distance of 20 cm from the centre of the lens on the axis.

48.6 Which lens has the greater optical power: a lens of focal length 10 cm or one with a focal length of 50 cm?

Working out the lens equation: We use similar triangles on the diagram of page 144 to form two equations for O/I.

Using the ray through the middle of the lens we know that:

$$\frac{O}{I} = \frac{u}{v}$$

Using the other diagonal ray, and the two triangles it forms, we can also write:

$$\frac{O}{I} = \frac{f}{v - f}$$

Equating the two expressions:

$$\frac{u}{v} = \frac{f}{v - f} \implies \frac{v}{u} = \frac{v - f}{f} \implies \frac{v}{u} = \frac{v}{f} - 1$$

$$\implies \frac{1}{u} = \frac{1}{f} - \frac{1}{v} \quad \text{or} \quad \frac{1}{v} = \frac{1}{f} - \frac{1}{u}$$

This can be worked out more easily on a calculator like this:

$$v^{-1} = f^{-1} - u^{-1}, \text{ so } v = (f^{-1} - u^{-1})^{-1}.$$

Example 2 – Calculate the image distance (v) of a 5.0 D lens, if the object distance (u) is 30 cm.

$$P = 1/f, \text{ so} f = 1/P = 1/5.0 = 0.2 \text{ m}$$

$$1/v = 1/f - 1/u = 1/0.2 - 1/0.3 = 1.667,$$

$$\text{so } v = 1/1.667 = 0.6 \text{ m} = 60 \text{ cm}$$

48.7   Remember that $\frac{1}{4}$ and $1 \div 4$ and 4^{-1} all mean the same thing. Use the x^{-1} button on your calculator to calculate these values.

(a) 2^{-1} (c) 0.5^{-1} (e) $\frac{1}{50}$

(b) 4^{-1} (d) 20^{-1} (f) $1 \div 7$

48.8 Calculate the following using the x^{-1} button on your calculator.

(a) $3^{-1} + 4^{-1}$ (c) $(2^{-1} - 4^{-1})^{-1}$ (e) $\left(\frac{1}{3} - \frac{1}{12}\right)^{-1}$

(b) $2^{-1} - 4^{-1}$ (d) $(4^{-1} + 12^{-1})^{-1}$ (f) $\frac{1}{(1/2 - 1/5)}$

48.9 Use the formula to work out the image distance for each situation.
Assume that the focal length of the lens is 5.0 cm.

(a) Object distance = 20 cm (d) Object distance = 7.0 cm

(b) Object distance = 10 cm (e) Object distance = 1.0 cm

(c) Object distance = 15 cm (f) Object distance = 4.0 cm

In the final two cases, v is negative (the image is to the left of the lens). This is a virtual image, meaning that no rays actually cross there. A screen at that position would not show a bright spot.

$$\text{magnification} = \text{image height / object height} = \frac{I}{O} = \frac{v}{u}$$

A magnification of 2.0 means that the image is twice as tall as the object. Magnification of 1.0 gives an image the same height as the object. The magnification number does not say whether the image is 'upside down'.

48.10 Work out the image distance and magnification for the following convex lenses, and state whether the image is real or virtual.

(a) $f = 10$ cm, object distance = 20 cm

(b) $P = 4.0$ D, object distance = 10 cm

(c) $f = 50$ cm, object distance = 75 cm

(d) $P = 5.0$ D, object distance = 4.0 cm

$^{30}/_{39}$

A convex lens makes a virtual image if the object distance < focal length $(u < f)$.

A convex lens makes a real, magnified image if $f <$ object distance $< 2f$.

A convex lens makes a real, diminished image if $u > 2f$.

49   Concave Lenses ♡

In the diagram below, the object has size O, the image size I, and the lens has a focal length f. The lens now causes the rays to diverge. We can work out the location of the image by drawing two rays through the system.

1. A ray passing through the centre of the lens that does not bend.

2. A ray travelling parallel to the axis will bend at the lens so that it *appears* to come from the focal point F (distance f from the lens). On the diagram, we draw a dashed line from F to the lens, and a solid line from there on (to denote an actual ray.

3. The image I is drawn where the lines cross.

4. The object distance is labelled u. The image distance is labelled v.

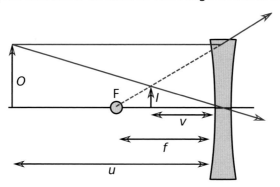

49.1   Copy and complete the ray diagrams by drawing the path of each ray of light after passing through the lens and hence find the position, size and orientation of the image. State whether the image formed is real or virtual.

The power formula $P = 1/f$ is used for concave lenses, just as it is for convex lenses. However concave lens powers are negative. When giving the power of a lens, always give the sign to make it clear whether you mean a convex or concave lens.

Example 1 – Calculate the power of a concave lens with a focal length of 4.0 cm.

$$4.0 \text{ cm} = 0.04 \text{ m} \quad P = 1/f = 1/0.040 = 25 \text{ D}$$

This is a concave lens, so we use a negative power $P = -25$ D.

Example 2 – Calculate the focal length of a -0.8 D lens. The power is negative, so this is a concave lens.

$$P = 1/f, \text{ so } f = 1/P = 1/(0.8 \text{ D}) = 1.25 \text{ m}$$

49.2 Use the equation $P = f^{-1}$ to work out the power in dioptres of lenses with the following focal lengths:
    (a) convex, $f = 1.6$ m.   (c) convex, $f = 5.0$ cm.
    (b) concave , $f = 2.0$ m.   (d) concave, $f = 7.0$ cm.

49.3 Calculate the focal length, and say whether it is concave or convex.
    (a) A -2.5 D lens.   (c) A $+20$ D lens.
    (b) A -1.5 D lens.   (d) A -40 D lens.

We use similar triangles in the diagram of page 148 to form two equations for O/I. The ray through the middle of the lens yields:

$$\frac{O}{I} = \frac{u}{v}$$

Using the line from F to the lens via the top of I, we can also write:

$$\frac{O}{I} = \frac{f}{f - v}$$

Equating the two expressions:

$$\frac{u}{v} = \frac{f}{f-v} \Rightarrow \frac{v}{u} = \frac{f-v}{f} \Rightarrow \frac{v}{u} = 1 - \frac{v}{f}$$

$$\Rightarrow \frac{1}{u} = \frac{1}{v} - \frac{1}{f} \Rightarrow \frac{1}{v} = \frac{1}{u} + \frac{1}{f}$$

This is the same as the equation on page 146 for a convex lens, if you flip the signs of f and v. Making v negative makes sense given that our image is to the left of the lens. Making f negative also makes sense as we remember that $P = 1/f$ and concave lenses have negative powers.

- For all lenses $1/v = 1/f - 1/u$, where

- a negative v means that the image is to the left of the lens, and

- a negative value of f means that the lens is concave.

49.4 Use a scale diagram, or the formula, to work out the image distance for each situation below. Assume that you have a 5.0 cm concave lens ($f = -5.0$ cm), and that the object height is 8.0 cm.

(a) Object distance = 20 cm   (c) Object distance = 15 cm

(b) Object distance = 10 cm   (d) Object distance = 4.0 cm

As with convex lenses, the magnification = $I/O = v/u$, where we ignore the sign of v.

49.5 Work out the image distance and magnification for the following lenses, and state whether the image is real or virtual.

(a) A concave lens with 10 cm focal length, 20 cm object distance.

(b) A concave lens with 20 cm focal length, 10 cm object distance.

(c) A convex lens with 30 cm focal length, 60 cm object distance.

(d) A convex lens with 40 cm focal length, 30 cm object distance.

(e) A convex lens with 50 cm focal length, 75 cm object distance.

(f) A convex lens with 60 cm focal length, 1.5 m object distance.

A concave lens makes a virtual, diminished image.

50   Intensity and Radiation ♡

The intensity of light, sound or other radiation depends on the

- power of the wave, and

- the size of the area in which the waves are focused.

Formula:

$$\text{intensity (W/m}^2) = \text{power (W)}/\text{area (m}^2) \qquad I = P/A$$

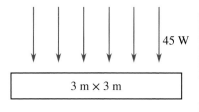

Example 1
Intensity $= P/A = 45\,\text{W} \div 9\,\text{m}^2$
$= 5\,\text{W/m}^2$

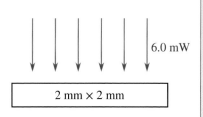

Example 2
Area $= 2\,\text{mm} \times 2\,\text{mm} = 0.002\,\text{m} \times$
$0.002\,\text{m} = 4 \times 10^{-6}\,\text{m}^2$
Intensity $= P/A$
$= (6 \times 10^{-3}\,\text{W}) \div (4 \times 10^{-6}\,\text{m}^2)$
$= 1.5 \times 10^3\,\text{W/m}^2$
$= 1\,500\,\text{W/m}^2$

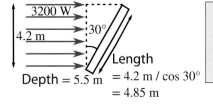

Example 3
Area lit $= 5.5\,\text{m} \times 4.85\,\text{m} = 26.7\,\text{m}^2$
Intensity $= P/A$
$= 3\,100\,\text{W} \div 26.7\,\text{m}^2 = 120\,\text{W/m}^2$

Point Sources
To work out the intensity at a distance from a point source, we imagine it
shining light in all directions, making the shape of a sphere.

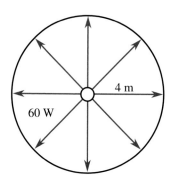

Intensity 4 m from the source
= power / area illuminated
= power / surface area of a 4 m sphere
= $P/(4\pi r^2)$
= $60/(4\pi \times 4^2) = 60/201 = 0.30$ W/m^2.

50.1 A light bulb radiates at 60 W (thermal and light) evenly in all directions. What is the intensity if

(a) this light all falls in a 5.0 m^2 area?

(b) the light all falls on a 10 m^2 area?

How much area would the bulb light if it were placed

(c) in the middle of a spherical room of radius 3.0 m?

(d) in the middle of a spherical room of radius 6.0 m?

What would the intensity be at the walls of

(e) the spherical room in question (c)?

(f) the spherical room in question (d)?

50.2 A car has 50 W headlamps on it.

(a) Calculate the intensity you would expect from a single headlamp bulb at a distance of 400 m if it shone light in all directions
equally.

(b) In practice the intensity 400 m from a headlamp is much higher.
Why?

50.3 Calculate the intensity you would expect from a 1.0 W torch bulb
at a distance of 3.0 m.

50.4 **PLEASE _DON'T_ DO THIS AT HOME** – you might damage your eyes. If you hold a 100 W bulb about 7.0 cm from your eye, it looks as bright as the Sun. We shall use this fact to calculate the power of the Sun.

(a) What is the intensity of the light 7.0 cm from a 100 W bulb?

(b) What is the intensity of sunlight at the surface of the Earth? [no calculation needed]

(c) The Sun is 1.5×10^{11} m from the Earth. Calculate the surface area of a sphere with this radius.

(d) Use your answers to (b) and (c) to determine the power output (luminosity) of the Sun in watts.

50.5 You want to make a solar power station giving an output of 2 GW (2×10^9 W). Use your answer to Q50.4b to calculate:

(a) The ground area needed for solar cells if they are 100% efficient.

(b) The ground area needed for solar cells if they are 20% efficient.

50.6 Fill in the blanks using the words at the end.

Any energy given off in the form of waves can be called _____. In this sense, _____, radio masts and oven _____ all give off radiation. However, none of these have the ability to _____ atoms - to temporarily change the number of _____ they carry, and thus cause them to act strangely in _____ reactions. Ionizing radiation is either conventional _____ radiation of very high _____ (UV light, X-rays or gamma rays) or a stream of charged particles (like alpha or beta). If your cells receive too much ionizing radiation, a _____ may occur. This may be harmless, it might cause the cell to die, it might prevent the cell _____, or it could cause the cell to reproduce uncontrollably. This last possibility is the root of many _____. Other effects of exposure to ionizing radiation include skin burns, nausea, destruction of _____, hair loss, and sterility. At exceptionally high doses, the thermal energy given to the cells by the ionizing radiation can prove instantly fatal. **Words**: cancers, grills, bone marrow, frequency, electrons, electromagnetic, mutation, radiation, ionize, mobile phones, chemical, reproducing.

Nuclear

All matter is made up of atoms.

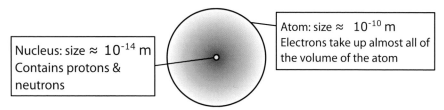

Nucleus: size $\approx 10^{-14}$ m
Contains protons & neutrons

Atom: size $\approx 10^{-10}$ m
Electrons take up almost all of the volume of the atom

Particles:

Name	Symbol	Relative charge	Relative mass
Proton	^1_1p or ^1_1H	$+1$	1.0000
Electron[1]	$^0_{-1}e$ or $^0_{-1}\beta$	-1	0.0005
Neutron	^1_0n	0	1.0016
Positron	^0_1e or $^0_1\beta$	$+1$	0.0005

No internal structure inside an electron has been found; it is a fundamental particle.

Every particle has an anti-particle of opposite charge but identical mass. The anti-electron is called a positron. If a particle meets its antiparticle, the two annihiliate each other, and their energy is given out as gamma rays.

The atomic number, Z, is the number of protons in a nucleus.

The mass number, A, is the number of protons plus neutrons in a nucleus.

$^{14}_6C$ (also written as carbon-14) is an isotope of carbon with a mass number of $A = 14$. It has $Z = 6$ protons, 6 electrons and $14 - 6 = 8 (= A - Z)$ neutrons.

All atoms with the same number of protons belong to the same element. They will behave identically in any chemical process.

---

[1]Beta (β^-) radiation consists of free electrons moving very quickly. Beta particles are electrons emitted from nuclei- so not all electrons are beta particles.

Two atoms are said to be isotopes of the same element if they have the same number of protons but different numbers of neutrons. They will, consequently, have different masses.

51.1 (a) How many protons are there in a helium atom?

Element	Z
Hydrogen (H)	1
Helium (He)	2
Lithium (Li)	3
Nitrogen (N)	7
Oxygen (O)	8
Uranium (U)	92

(b) How many electrons are there in a uranium atom?

(c) How many protons are there in a lithium−7 atom?

(d) How many neutrons are there in a lithium−7 atom?

51.2 For the atoms in the table, fill in the number of protons, neutrons and electrons they have.

Atom	Protons	Neutrons	Electrons
Uranium−238	(a)	(b)	(c)
Oxygen−16	(d)	(e)	(f)
^{3}He	(g)	(h)	(i)
^{235}U	(j)	(k)	(l)

51.3 State the number of protons, neutrons and electrons in a $^{63}_{29}$Cu$^+$ ion.

♡ Protons and neutrons are made of quarks. Up quarks (u) have charge $+2/3$, while down quarks (d) have charge $-1/3$.

51.4 There are three quarks in a proton. How many of them are up quarks, and how many are down quarks?

51.5 What is a neutron is made of?

51.6 During beta minus decay, a neutron turns into a proton. What happens in terms of quarks?

51.7 During beta plus decay, a proton turns into a neutron. What happens in terms of quarks?

$^{15}/_{20}$

52   Radioactive Decay ♡

Some nuclei are stable, and will remain as they are for ever. Others are unstable. After an unpredictable period of time, unstable nuclei will change. This change is called decay. When a nucleus decays, it gives out highly energetic, ionizing radiation. The main forms of ionizing radiation are alpha particles, beta particles and gamma rays.

Type of decay	Particle given out	Penetrating ability	Ionising ability	Change to the original nucleus
Alpha	$^4_2\alpha$ - Helium nucleus (2 protons + 2 neutrons)	Low - stopped by 5 cm of air, by skin or paper	High	Mass number reduces by 4 Atomic number reduces by 2
Beta minus	$^0_{-1}\beta$ - High speed electron produced when a neutron turns into a proton	Medium - can pass 1 mm of aluminium, but stopped by 2 cm	Medium	Mass number doesn't change Atomic number increases by 1
Beta plus	$^0_{+1}\beta$ - High speed positron (anti-electron) produced when a proton turns into a neutron	Very low - annihilates on contact with normal matter	N/A	Mass number doesn't change Atomic number reduces by 1
Gamma	$^0_0\gamma$ - High frequency electromagnetic wave	High - can pass through many cm of lead	Low	Mass number doesn't change Atomic number doesn't change Excess nuclear potential energy is released

Example 1 - Write the equation for the alpha decay of $^{241}_{95}$Am into Np.

The symbol for the alpha particle is $^{4}_{2}\alpha$.
We write the equation $^{241}_{95}$Am $\longrightarrow$ Np + $^{4}_{2}\alpha$ to show the decay.

Next, we need to put mass and atomic numbers on the Np. We do this using the rules in the table: $^{241}_{95}$Am $\longrightarrow$ $^{237}_{93}$Np + $^{4}_{2}\alpha$.

Notice that once the equation is complete the numbers on the top balance ($214 = 237 + 4$), as do the numbers on the bottom ($95 = 93 + 2$).

Example 2 - Write the equation for the beta minus decay of $^{3}_{1}$H into He.

Firstly, we write $^{3}_{1}$H $\longrightarrow$ He + $^{0}_{-1}\beta$, then put numbers on He to balance it: $^{3}_{1}$H $\longrightarrow$ $^{3}_{2}$He + $^{0}_{-1}\beta$.

Again notice that the top row balances ($3 = 3 + 0$) and so does the bottom ($1 = 2 - 1$).

Write equations for the following decays.

52.1 The alpha decay of $^{238}_{92}$U into Th.

52.2 The beta minus decay of $^{14}_{6}$C into N.

52.3 The gamma decay of $^{60}_{27}$Co. *[Hint: with no change to the atomic number, the decay produces Co]*

52.4 The beta minus decay of $^{90}_{38}$Sr into Y.

52.5 ♡ The beta plus decay of $^{11}_{6}$C into B.

52.6 The beta minus decay of $^{8}_{3}$Li into Be.

52.7 The beta minus decay of $^{40}_{19}$K into Ca.

52.8 The alpha decay of $^{239}_{94}$Pu into U.

52.9 The alpha decay of $^{210}_{86}$Rn into Po.

52.10 ♡ The beta plus decay of $^{14}_{8}$O into N.

53   Half-Life

Nuclear decay is random. You can not predict when an individual nucleus will decay. However, if you have many millions of nuclei, you can make a good prediction of how many will decay in a certain amount of time.

The half-life is the average time taken for the number of unstable nuclei to halve.

The half-life is also the average time taken for the activity (number of decays each second) to halve.

> Example - The half-life of $^3_1$H is 12 years. A source starts with an activity of 150 Bq (150 decays per second). Estimate the activity 12 and 24 years after the start.
>
> After 12 years, one half-life has passed, so the activity will halve to 75 Bq. After 24 years, a second half-life has passed, halving the activity again to $75 \times 0.5 = 37.5$ Bq.

53.1 $^{14}_{6}$C has a half-life of 5 700 years. A sample is 5 700 years old and has an activity of 200 Bq.

  (a) What was the initial activity?

  (b) What will the activity be 5 700 years in the future?

53.2   A sample starts with 10^{16} nuclei of $^3_1$H, which has a half-life of 12 years.

  (a) How many $^3_1$H nuclei will this sample contain after 12 years?

  (b) How many $^3_1$H nuclei will this sample contain 24 years after the start?

  (c) How many $^3_1$H nuclei will this sample contain 36 years after the start?

53.3   These questions are about $^{13}_{7}$N, which has a half-life of 10 minutes.

  (a) If I start with 6 000 000 nuclei, how many will remain after 10 minutes?

(b) If the activity was 600 Bq initially, what will it be after 30 minutes?

(c) If the activity was 24 000 Bq initially, what will it be one hour later?

53.4 Suppose the activity of a sample of radioactive material was 100 Bq at the start. What would you divide 100 Bq by to obtain the activity

(a) 1 half-life after the start?

(b) 2 half-lives after the start?

(c) 3 half-lives after the start?

(d) 4 half-lives after the start?

(e) 20 half-lives after the start?

(f) n half-lives after the start?

53.5 ♡ Use your reasoning from Q53.4 to answer this. The half-life of $^{13}_{7}N$ is 10 minutes. The initial activity of a sample of $^{13}_{7}N$ is 100 Bq. Determine the activity 5 minutes later.

$^{12}/_{15}$

54   Fission – The Process

Nuclear fission is the process by which one atomic nucleus splits to form two atomic nuclei. If the nucleus that splits has an atomic number above 26 (the atomic number of iron, Fe), the nuclear reaction releases energy.

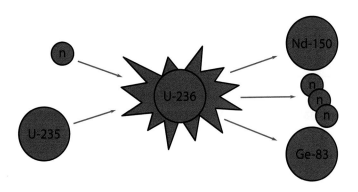

Heavy nuclei can often be made even less stable by absorbing an additional neutron. Uranium-235, for example, has 143 neutrons in the nucleus. If a uranium-235 nucleus absorbs a neutron, it quickly fissions (splits) into two daughter nuclei and a few free neutrons. The two daughter nuclei tend to have a mass ratio close to 2:3; however, this is random, and two or three free neutrons are also released. The free neutrons can hit other uranium nuclei, and could cause them to split. If these neutrons cause uranium nuclei to fission, releasing further neutrons which cause other uranium nuclei to fission, we call this a chain reaction.

The total number of neutrons before a fission is equal to the total number of neutrons after a fission.

Similarly, the total number of protons before a fission is equal to the total number of protons after a fission.[1]

The fission products are usually highly radioactive.

---

[1]For an interactive periodic table where you can check isotopes, masses, half lives etc, see www.ptable.com.

Example - Balanced nuclear fission reactions:

$$^{235}_{92}U + ^{1}_{0}n \longrightarrow ^{150}_{60}Nd + ^{83}_{32}Ge + 3\,^{1}_{0}n$$

Check mass (top) numbers balance: $235 + 1 = 150 + 83 + (3 \times 1)$

Check atomic (bottom) numbers balance: $92 + 0 = 60 + 32 + (3 \times 0)$

$$^{239}_{94}Pu + ^{1}_{0}n \longrightarrow ^{149}_{58}Ce + ^{88}_{36}Kr + 3\,^{1}_{0}n$$

54.1 Fill in the missing numbers in the following fission reactions:

(a) $^{235}_{92}U + ^{1}_{0}n \longrightarrow ^{196}_{77}Ir + ^{[\ ]}_{[\ ]}P + 2\,^{1}_{0}n$

(b) $^{235}_{92}U + ^{1}_{0}n \longrightarrow ^{167}_{66}Dy + ^{[\ ]}_{[\ ]}Fe + 3\,^{1}_{0}n$

(c) $^{235}_{92}U + ^{1}_{0}n \longrightarrow ^{167}_{65}Tb + ^{65}_{[\ ]}Co + [\ ]^{1}_{0}n$

(d) $^{239}_{[\ ]}Pu + ^{1}_{0}n \longrightarrow ^{155}_{62}Sm + ^{83}_{32}Ge + [\ ]^{1}_{0}n$

(e) $^{233}_{[\ ]}U + ^{1}_{0}n \longrightarrow ^{154}_{[\ ]}Pm + ^{[\ ]}_{[\ ]}Ga + 4\,^{1}_{0}n$

54.2 When a nucleus of uranium-235 captures a neutron, fission takes place. One possible fission is:

$$^{235}_{92}U + ^{1}_{0}n \longrightarrow ^{90}_{36}Kr + ^{144}_{56}Ba + [\ ]^{1}_{0}n$$

Calculate how many neutrons are released.

54.3 When a nucleus of uranium-235 captures a neutron, fission takes place. One possible fission is:

$$^{235}_{92}U + ^{1}_{0}n \longrightarrow ^{95}_{36}Kr + ^{x}_{y}Ba + 3\,^{1}_{0}n$$

Calculate x and y.

54.4 When a nucleus of plutonium-239 captures a neutron, fission takes place. One possible fission is:

$$^{239}_{94}Pu + ^{1}_{0}n \longrightarrow ^{137}_{52}Te + ^{x}_{y}Z + 3\,^{1}_{0}n$$

Calculate x and y and identify the element Z.

54.5 State the two most viable fuels for nuclear fission reactions.

54.6 Explain why there is so much variety in the daughter isotopes produced in fission reactions and why this adds to the challenge of managing nuclear waste.

54.7 For a sustainable chain reaction, one neutron released in the reaction must go on to cause one further fission reaction. Describe what happens to the other free neutrons.

54.8 In the reaction equations in this section, an intermediate step has been missed off (for example, in Q54.1a, $^{235}_{92}U$ plus one neutron becomes $^{236}_{92}U$ before the fission reaction takes place). How can we justify ignoring this middle step?

54.9 In the following examples of nuclear disintegrations, identify the missing numbers, elements or particles.

(a) $^{238}_{92}U \longrightarrow ^{[\ ]}_{[\ ]}Th + ^{4}_{2}He$

(b) $^{234}_{90}Th \longrightarrow ^{[\ ]}_{[\ ]}Pa + ^{0}_{-1}e$

(c) $^{234}_{91}Pa \longrightarrow ^{[\ ]}_{[\ ]}[\ ] + ^{0}_{-1}e$

(d) $^{222}_{86}Rn \longrightarrow ^{[\ ]}_{[\ ]}Po + ^{4}_{2}He$

(e) $^{[\ ]}_{82}Pb \longrightarrow ^{210}_{[\ ]}Hg + ^{[\ ]}_{2}He$

(f) $^{14}_{6}C \longrightarrow ^{[\ ]}_{[\ ]}N + ^{0}_{-1}e$

(g) $^{[\ ]}_{82}Pb \longrightarrow ^{214}_{[\ ]}Bi + ^{[\ ]}_{-1}e$

54.10 Fill in the missing numbers and symbols in the following nuclear processes.

(a) $^{[\ ]}_{[\ ]}Si \longrightarrow ^{28}_{14}Si + ^{1}_{0}n$

(b) $^{[\ ]}_{[\ ]}K \longrightarrow ^{40}_{18}Ar + ^{1}_{1}H$

(c) $^{52}_{24}Cr \longrightarrow ^{48}_{22}Ti + ^{[\ ]}_{[\ ]}[\ ]$

(d) $^{[\ ]}_{[\ ]}Cr \longrightarrow ^{55}_{25}Mn + ^{0}_{-1}\beta$

(e) $^{124}_{57}La + ^{1}_{0}n \longrightarrow ^{40}_{20}Ca + ^{[\ ]}_{[\ ]}Rb$

(f) $^{142}_{62}Sm + ^{1}_{1}H \longrightarrow ^{55}_{25}Mn + ^{[\ ]}_{[\ ]}Sr$

(g) $^{113}_{49}In + ^{[\ ]}_{[\ ]}[\ ] \longrightarrow ^{9}_{4}Be + ^{108}_{47}Ag$

(h) $^{73}_{32}Ge + ^{19}_{9}F \longrightarrow ^{92}_{42}Mo + ^{[\ ]}_{[\ ]}[\ ]$

55   Fission – The Reactor

Nuclear fission reactors convert nuclear energy to thermal energy. The nuclear energy is locked away in the nuclei of atoms with large atomic masses. The most common nuclear fuel is uranium−235. When the nucleus of a uranium−235 atom fissions, it becomes two smaller nuclei plus two or three free neutrons.

Control rods - inserting them deeper between the fuel roads decreases the reaction rate.

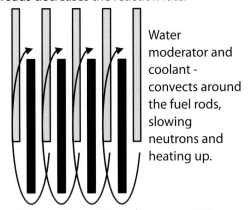

Water moderator and coolant - convects around the fuel rods, slowing neutrons and heating up.

Fuel rods - contain uranium-235 and uranium-238. Enriched fuels contain a greater proportion of uranium-235.

The neutrons that are released from a fission reaction are too fast to be absorbed by other uranium−235 nuclei. A moderator, such as water or graphite, is used. This slows the neutrons (reducing their kinetic energy). The moderator is warmed and a coolant carries the thermal energy away. If water is used as a moderator, the water itself can be the coolant.

If one spare neutron from each fission reaction is slowed down enough and absorbed by another uranium−235 nucleus, the reaction is a self-sustaining chain reaction. If too many neutrons are absorbed, the reaction rate can exponentially grow - this is what happens when a nuclear fission bomb is detonated. To prevent the reaction rate increasing, control rods made from boron or cadmium are included in the reactor to absorb spare free neutrons.

The nuclear fuel rods, moderator, coolant and control rods are all in the nuclear reactor core, which is contained in a concrete domed building. Heat exchangers carry the energy out of the core.

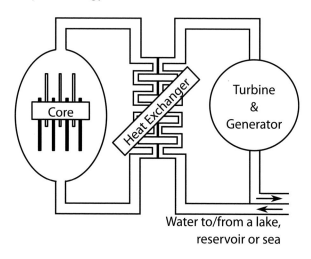

Water to/from a lake,
reservoir or sea

55.1 What is the function of
(a) the nuclear fuel?   (c) the moderator?
(b) the control rods?   (d) the coolant?
(e) the concrete containment structure?
(f) the heat exchanger?

55.2 Explain why is it necessary to slow down the free neutrons that are emitted from a fission reaction.

55.3 Describe what steps must be taken to ensure the chain reaction is self-sustaining.

55.4 Explain why the temperature within the reactor core must be closely monitored.

55.5 What safety mechanisms are in place in case the reaction rate starts to increase exponentially?

56   Energy from the Nucleus – Radioactivity & Fission ♡

You can calculate the energy released by a nuclear process if you know the total mass of the nuclei present before and the total mass of those present after the process.

$$\text{energy} = \text{change of mass} \times (\text{speed of light})^2 \qquad E = mc^2$$

In this equation, the mass is measured in kilograms, the energy is measured in joules and the speed of light is 3.00×10^8 m/s.

In a nuclear reaction, the products (once they have slowed to normal speeds) have less mass in total than the reactants had: Although mass-energy is conserved, the energy released by the reaction has a mass equivalent.

$$\text{energy released} = \text{mass 'lost'} \times (\text{speed of light})^2$$

Example – Calculate the energy released during the reaction:

$$^{241}_{95}\text{Am} \longrightarrow {}^{237}_{93}\text{Np} + {}^{4}_{2}\alpha$$

The masses of the nuclei are given in the table below.

^{241}Am	$4.001\,98 \times 10^{-25}$ kg
^{237}Np	$3.935\,43 \times 10^{-25}$ kg
α	6.645×10^{-27} kg

$$\text{Mass of reactants} = \text{mass of } {}^{241}\text{Am} = 4.001\,98 \times 10^{-25} \text{ kg}$$

$$\text{Mass of products} = \text{mass of } {}^{237}\text{Np} + \text{mass of } \alpha$$

$$= 3.935\,43 \times 10^{-25} \text{ kg} + 6.645 \times 10^{-27} \text{ kg}$$

$$= 4.001\,88 \times 10^{-25} \text{ kg}$$

$$\text{Difference in masses} = 4.001\,98 \times 10^{-25} \text{ kg} - 4.001\,88 \times 10^{-25} \text{ kg}$$

$$= 0.000\,10 \times 10^{-25} \text{ kg} \equiv 1.0 \times 10^{-29} \text{ kg}$$

$$\text{Energy released} = mc^2 = 1.0 \times 10^{-29} \text{ kg} \times (3.00 \times 10^8)^2$$

$$= 9.0 \times 10^{-13} \text{ J}$$

This may seem a very small amount per reaction, but it is over 5 000 000 times larger than the energy given out in chemical reactions.

Some masses of nuclei for use in the questions:

$^{1}_{0}$n	$0.016\,749 \times 10^{-25}$ kg	α	6.645×10^{-27} kg
$^{87}_{35}$Br	$1.443\,031 \times 10^{-25}$ kg	$^{103}_{40}$Zr	$1.708\,773 \times 10^{-25}$ kg
$^{134}_{54}$Xe	$2.223\,061 \times 10^{-25}$ kg	$^{147}_{57}$La	$2.439\,291 \times 10^{-25}$ kg
$^{189}_{81}$Tl	$3.137\,255 \times 10^{-25}$ kg	$^{193}_{83}$Bi	$3.203\,808 \times 10^{-25}$ kg
$^{206}_{82}$Pb	$3.419\,541 \times 10^{-25}$ kg	$^{206}_{84}$Po	$3.419\,623 \times 10^{-25}$ kg
$^{210}_{84}$Po	$3.486\,084 \times 10^{-25}$ kg	$^{210}_{86}$Rn	$3.486\,179 \times 10^{-25}$ kg
$^{212}_{83}$Bi	$3.519\,444 \times 10^{-25}$ kg	$^{216}_{85}$At	$3.586\,032 \times 10^{-25}$ kg
$^{234}_{90}$Th	$3.885\,568 \times 10^{-25}$ kg	$^{235}_{92}$U	$3.902\,162 \times 10^{-25}$ kg
$^{238}_{92}$U	$3.952\,090 \times 10^{-25}$ kg	$^{239}_{94}$Pu	$3.968\,700 \times 10^{-25}$ kg

56.1   Using the previous information calculate the energy released during the alpha decays of:

(a) $^{193}_{83}$Bi

(b) $^{210}_{84}$Po

(c) $^{216}_{85}$At

(d) $^{210}_{86}$Rn

(e) $^{238}_{92}$U

[Hint: use the method in the Section 56, Example 1]

56.2   When a uranium nucleus fissions, there are various products which can be made. One typical reaction is

$$^{235}_{92}U + ^{1}_{0}n \longrightarrow ^{147}_{57}La + ^{87}_{35}Br + 2^{1}_{0}n$$

(a) Calculate the total mass of the reactants.

(b) Calculate the total mass of the products.

(c) The mass 'lost' is the energy lost to the nuclei. This energy is released in the form of kinetic energy. Calculate the mass lost.

(d) Use the equation $E = mc^2$ to work out how much energy has been lost from the nuclei (and gained in kinetic energy).

(e) The energy you calculated in (d) was released when one nucleus of uranium was fissioned. Use the mass of this nucleus to work out how much energy you could get out of 1.0 kg of uranium if you fissioned all of the nuclei.

(f) A nuclear power station has a thermal power output of 3.0×10^9 W. Calculate how much energy is generated in one year of continuous operation.

(g) Use your answers to 56.1(e) and (f) calculate the minimum amount of uranium you would need to fuel the power station for a year.

(h) Why must the reaction make at least two neutrons?

(i) The combustion of one carbon atom (mass $= 2.0 \times 10^{-26}$ kg) releases 6.6×10^{-20} J of energy. Calculate the mass of carbon (e.g. coal) which would need to be burnt each day to have the same thermal power output as the 3.0×10^9 W nuclear station.

(j) The mass of high-level nuclear waste produced per year by the nuclear power station will be similar to your answer to (g). If the material had a density of 6000 kg/m^3, work out the volume of nuclear waste produced in one year's operation.

(k) A cube-shaped underground chamber is to hold the waste produced by the power station over its 20 year operating lifetime. Use your answer to (j) to work out the side length of the cube.

(l) Since fission products absorb too many neutrons, nuclear fuel has to be removed from a reactor when only 5% of the fissile uranium has been used. Repeat part (k) on the assumption that a country does not 'reprocess' its nuclear fuel to remove the fission products, and that fuel is thrown away when only 5% of it has been used up.

(m) Work out the mass of the waste products of burning the coal in (i). When you burn 12 g of carbon, you need 32 g of oxygen, and so for every 12 g of coal burnt, 44 g of waste is made.

(n) Work out the volume of oxygen required to burn the coal in (i). Assume that every 12 g of carbon needs 32 g of oxygen to burn, and that 32 g of oxygen gas has a volume of 0.024 m^3.

56.3   A Plutonium-239 fission is by the following reaction:

$$^{239}_{94}\text{Pu} + ^{1}_{0}\text{n} \longrightarrow ^{134}_{54}\text{Xe} + ^{103}_{40}\text{Zr} + x^{1}_{0}\text{n}$$

Give the value of x, on the right hand side of this equation, and calculate the kinetic energy of the products.

57   Fusion – The Process ♡

8/10

Nuclear fusion is the process by which two atomic nuclei combine to form a single atomic nucleus. If the two nuclei that go into a fusion reaction have an atomic number below 26 (the atomic number of iron), the nuclear reaction can release energy.

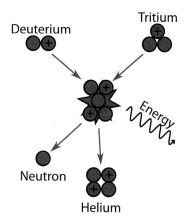

Atomic nuclei are positively charged. Like charges repel each other. The strength of the electrostatic repulsive force increases as the distance between the nuclei decreases. This force prevents the two nuclei getting close enough to fuse unless the nuclei are moving very fast. If the nuclei are moving very fast, the electrostatic repulsive force cannot stop two nuclei moving towards each other until it is too late; they are close enough to fuse, under the action of the strong nuclear force, and become a single nucleus. This barrier to nuclear fusion is called the Coulomb barrier.

Once the Coulomb barrier is breached, far more energy is released than the energy required to breach the Coulomb barrier in the first place.

In stars, atomic nuclei are given sufficient energy to breach the Coulomb barrier because the temperature in the star is so high. On Earth, experimental fusion reactor designs focus on energy-efficient ways of increasing the speed of the atomic nuclei and on novel methods for reducing the Coulomb barrier. To date, no fusion reactor on Earth has released energy at a self-sustaining rate for more than a fraction of a second.

The best atomic nuclei to use as a fuel in a fusion reactor are hydrogen, which has the fewest number of protons of any atomic nucleus. The Coulomb barrier can be more easily overcome by using isotopes of hydrogen that contain neutrons, such as deuterium (one proton, one neutron) and tritium (one proton, two neutrons).

Hydrogen is a readily available fuel because it is present in water, which covers approximately 70% of the Earth's surface.

57.1 When two hydrogen atoms fuse, what element is produced?

57.2 In the Sun, helium−4 atoms are produced. The only atomic nuclei that go into the reaction are hydrogen−1. The reaction has many stages, the last of which produces two protons and one helium−4 nucleus.

   (a) How many nucleons are input into the reaction chain?

   (b) All of these nucleons are initially protons, but some of them can transform into neutrons. How many protons transform into neutrons for each occurrence of this reaction chain?

   (c) Each proton that transforms into a neutron emits a particle with positive charge. What is this particle called?

   (d) What happens to the emitted particle in (c)?

57.3 At what temperature can nuclear fusion of hydrogen naturally occur?

57.4 In the early 1950s, the first nuclear fusion device was successfully tested. What was the purpose of this device?

57.5 Stars larger than the Sun involve another nuclear fusion chain, called the CNO cycle. What do the letters C, N and O stand for?

57.6 In what form is the energy released from a nuclear fusion reaction initially?

57.7 Explain why the mass of the Sun reduces by about 4 000 000 tonnes per second. [See Q50.4d for luminosity.] 1 tonne = 1000 kg

58   Energy from the Nucleus – Fusion ♡

The methods needed for working out the energy released are explained in full on Energy from the Nucleus – Radioactivity & Fission ♡ - P165.

The most promising fusion reaction, as far as power stations are concerned, is this:

$$^2_1H + {}^3_1H \longrightarrow {}^4_2He + {}^1_0n$$

The masses of some nuclei are given in the table below:

^1_0n	1.6749×10^{-27} kg	^2_1H	3.3436×10^{-27} kg
^3_1H	5.0074×10^{-27} kg	^4_2He	6.6447×10^{-27} kg

58.1   Consider the reaction $^2_1H + {}^3_1H \longrightarrow {}^4_2He + {}^1_0n$.

(a) Calculate the total mass of the reactants.

(b) Calculate the total mass of the products.

(c) The mass 'lost' is the energy lost to the nuclei. This energy is released in the form of kinetic energy. Calculate the lost mass.

(d) Use the equation $E = mc^2$ to work out how much energy has been lost from the nuclei (and gained in kinetic energy). Take $c = 3.00 \times 10^8$ m/s.

(e) A 1.0 kg sample of fusion fuel contains equal numbers of ^2H and ^3H nuclei. How many ^2H nuclei would the sample contain? Ignore the mass of the electrons which would also be in the fuel.

(f) The energy you calculated in (d) was released when one fusion occurred. Use your answers to the previous questions to work out how much energy you could get out of 1.0 kg of fusion fuel (with ^2H and ^3H in equal numbers) if you fused it all.

(g) A nuclear power station has a power output of 3.0×10^9 W. Calculate how much energy is generated in one year of continuous operation.

(h) Use your answers to (f) and (g) to calculate how many kilograms of fusion fuel you need to fuel the power station for a year.

To answer the next two questions, you will need your answers to the Energy from the Nucleus – Radioactivity & Fission ♡ worksheet questions, P166.

58.2 Compare your answers to 56.2d and 58.1d to find out which produces more energy – one fission or one fusion.

58.3 Compare your answers to 56.2e and 58.1f to find out which produces more energy per kilogram of fuel – fission or fusion.

58.4 Why is radioactive waste not as big a problem with fusion as it is with fission?

$^{0}_{1}e$	$9.109\,4 \times 10^{-31}$ kg	$^{12}_{6}C$	$1.992\,10 \times 10^{-26}$ kg
$^{1}_{0}n$	$1.674\,9 \times 10^{-27}$ kg	$^{14}_{7}N$	$2.324\,63 \times 10^{-26}$ kg
$^{1}_{1}H$	$1.672\,6 \times 10^{-27}$ kg	$^{15}_{8}O$	$2.490\,59 \times 10^{-26}$ kg
$^{2}_{1}H$	$3.343\,6 \times 10^{-27}$ kg	$^{24}_{12}Mg$	$3.981\,72 \times 10^{-26}$ kg
$^{3}_{1}H$	$5.007\,4 \times 10^{-27}$ kg	$^{56}_{26}Fe$	$9.285\,85 \times 10^{-26}$ kg
$^{3}_{2}He$	$5.006\,4 \times 10^{-27}$ kg	$^{111}_{52}Te$	$18.414\,05 \times 10^{-26}$ kg
$^{4}_{2}He$	$6.644\,7 \times 10^{-27}$ kg		

58.5 Using the nuclear masses above, calculate the energy released in each of the following fusion reactions.

(a) $^{2}_{1}H + ^{2}_{1}H \longrightarrow ^{3}_{2}He + ^{1}_{0}n$

(b) $^{3}_{1}H + ^{3}_{1}H \longrightarrow ^{4}_{2}He + 2\,^{1}_{0}n$

(c) $4\,^{1}_{1}H \longrightarrow ^{4}_{2}He + 2\,^{0}_{1}e$

(d) $^{1}_{1}H + ^{14}_{7}N \longrightarrow ^{15}_{8}O$

(e) $^{12}_{6}C + ^{12}_{6}C \longrightarrow ^{24}_{12}Mg$

(f) $^{56}_{26}Fe + ^{56}_{26}Fe \longrightarrow ^{111}_{52}Te + ^{1}_{0}n$

(g) What is different about the last reaction?

$^{14}/_{18}$

Gas

Definition:

$$\text{pressure} = \frac{\text{force (N)}}{\text{area (m}^2)} \qquad P = \frac{F}{A}$$

The unit of pressure is the pascal (Pa). $1\ \text{Pa} = 1\ \text{N/m}^2$.
Atmospheric pressure is approximately $1.01 \times 10^5\ \text{Pa} = 101\ \text{kPa}$.

The pressure a gas exerts on a wall depends on

- how often molecules hit the wall. This depends on the

  - size of the container (if the container is longer, molecules will take longer to cross it, and each molecule will collide with the wall less often).
  - speed of the molecules (this depends on the temperature).
  - number of molecules in the container.

- the momentum change when each molecule hits the walls. This depends on the

  - mass of the molecules.
  - speed of the molecules (which depends on the temperature).

If the temperature is constant (which is usually the case if the gas is compressed or expands slowly), the speed of the molecules doesn't change. Halving the volume of the container doubles the gas pressure because each molecule only takes half the time to cross it – so hits the walls twice as often.

Equation for Boyle's Law (constant temperature)

$$\text{pressure} \times \text{volume} = \text{constant} \qquad p_1 V_1 = p_2 V_2$$

where $_1$ means 'before the change' and $_2$ means 'after the change'

Example – 40 cm^3 of gas at atmospheric pressure is squeezed into a volume of 10 cm^3. What is the new pressure?
$p_1V_1 = p_2V_2$, so 101 kPa × 40 cm^3 = p_2 × 10 cm^3, so 4 040 = 10p_2
p_2 = 4 040/10 = 404 kPa.

The average kinetic energy of molecules in a gas depends on temperature. The average kinetic energy of molecules is proportional to the temperature, if the temperature is measured in kelvins (K), where

temperature in kelvins (K) = temperature in degrees Celsius (°C) + 273.

The temperature of 0 K = −273 °C is called absolute zero. If you were able to cool a gas right down to this level, the molecules would be stationary. You couldn't cool it any further - this is the coldest temperature possible.

59.1 (a) What is 1.00 cm^2 in square metres?

(b) How much force does the atmosphere exert on a 1.00 cm^2 area?

59.2 A barometer measures the pressure of the atmosphere in millibar (mbar), where 1.0 mbar = 100 Pa. The surface area of the chamber in a barometer is 0.010 m^2, and the air pressure changes from 997 mbar to 1 013 mbar. What is the change in the force exerted by the air on the barometer's chamber?

59.3 Why does a gas exert a pressure on the walls of its container? Number the statements below to put them in the correct order to answer the question.

(a) bounce off with the same speed in a different direction.

(b) In a gas, molecules move around rapidly.

(c) By Newton's Third Law there must also have been a

(d) They frequently collide with the wall. When this happens they

(e) force exerted on the wall.

(f) their momentum must have changed, so

(g) there must have been a force on the molecule to cause this.

(h) This means that their velocity has changed and so

59.4 20 cm^3 of gas is at 100 kPa.

 (a) What will the pressure be on squeezing slowly down to 10 cm^3?

 (b) What will it be if the gas is allowed to expand slowly to 40 cm^3?

 (c) What will the volume be if the pressure is increased slowly to 1 000 kPa?

 (d) Suppose the change in (a) were done really quickly. What would the effect be on the speed of the molecules?

 (e) Would your final answer in (a) be lower or higher, if the change were instead done quickly?

59.5 The co-ordinates of a point A on the line of a pressure-volume graph constructed for a fixed mass of gas at constant temperature are $(40, 30)$. Point B $(60, y)$ and point C $(x, 10)$ also lie on the line. Calculate the values of x and y.

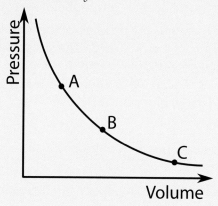

59.6 A certain car's suspension works by having a fixed mass of gas sealed inside a flexible capsule. Its pressure is usually 2.4×10^5 Pa and its volume is 2.0 litres. On a bumpy road, at one point, the gas inside the capsule is compressed to 1.5 litres. What is its pressure at this point? (Assume the gas temperature remains constant).

59.7 A child lets its helium-filled balloon float up, higher and higher into the air, becoming larger and larger. When it was at ground level, its volume was 4 000 cm^3 and the helium was at a pressure of 1.5×10^5 Pa.

What would the helium's pressure become if the volume increased to $6\,000$ cm^3 with no change of temperature?

59.8 In Boyle's Law, pressure and volume are inversely proportional ($p = k/V$, where k is a constant). What graph could you plot against p which would give a straight line? [*Hint: suppose we plotted p on the y-axis, we could plot some function of V on the x-axis.*]

59.9 Suppose the speed of each molecule doubled. What would happen to the pressure if the volume were fixed?

59.10 Convert the following temperatures into the other unit:

(a) 23 °C	(c) −50 °C	(e) 4 K	(g) 600 K
(b) 0 °C	(d) 90 K	(f) 0 K	(h) 6 000 °C

Additional Boyle's Law Questions

59.11 The average kinetic energy of molecules in air at 300 K is 6.21×10^{-21} J.

(a) What is the average kinetic energy of molecules in air at 600 K? *Hint: average kinetic energy is proportional to kelvin temperature. See page 7 if you need help with calculations involving proportionality.*

(b) What is the average kinetic energy of molecules in air at 373 K?

(c) What is the average kinetic energy of molecules in air at 0 °C?

(d) At atmospheric pressure, nitrogen liquefies at the temperature where the average kinetic energy of gas molecules would be 1.86×10^{-21} J. What is the boiling temperature of nitrogen? Give your answer in kelvin.

59.12 The mass of a nitrogen molecule is 4.65×10^{-26} kg. Use Q59.11(c) to calculate (a) the average value for (speed)^2 of a nitrogen molecule in air at 0 °C, and (b) a typical speed for such molecules.

59.13 The mass of a helium atom is 6.64×10^{-27} kg. Using the method of Q59.12, calculate a typical speed for helium atoms in air at 0 °C.

60   The Pressure Law ♡

In this situation, the volume is fixed (we use a rigid container). The gas is heated, and the pressure increases.

As the temperature of the gas goes up, the average speed and kinetic energy of the molecules increases.

This means that each second, more molecules hit each container wall, and also that on each collision there is a greater velocity (or momentum) change for the molecule, leading to a greater force on the wall.

The equation is

$$\frac{p_{\text{after}}}{T_{\text{after}}} = \frac{p_{\text{before}}}{T_{\text{before}}}$$

where T must be in kelvins.

Example – Starting with some gas at $20.0\,°C$ at a pressure of 101 kPa and heating it to $100\,°C$, what is the new pressure if the gas' volume is fixed?

1st stage: convert the temperatures to kelvins.

$$20.0\,°C + 273 = 293\text{ K} \qquad 100\,°C + 273 = 373\text{ K}$$

2nd stage: put the numbers into the equation.

$$\frac{p_{\text{after}}}{373\text{ K}} = \frac{101\text{ kPa}}{293\text{ K}}$$

3rd stage: rearrange the equation so that the thing you want to know is the subject, and calculate it.

$$p_{\text{after}} = 101\text{ kPa} \times \frac{373}{293} = 129\text{ kPa}$$

4th stage: put the temperatures back in $°C$ if necessary (not needed here).

60.1 I start with some gas at 30 °C at a pressure of 101 kPa and heat it to 200 °C. What will the new pressure be if I don't let the gas expand?

60.2 I start with some gas at −20 °C, at a pressure of 101 kPa, and heat it until the pressure is 202 kPa without letting it expand. What will the new temperature be?

60.3 A cylinder of compressed gas is at a temperature of 23 °C. It is cooled until it reaches a pressure of 2 000 kPa. It has to be cooled to 90 K before this happens. Calculate the starting pressure of the gas.

60.4 Work out the missing measurements from the following table, where each row is a separate question.

P_{before}	T_{before}	P_{after}	T_{after}
101 kPa	300 K	(a)	600 K
101 kPa	−23.0 °C	505 kPa	(b)
10.1 kPa	(c)	101 kPa	300 K
(d)	−183 °C	50.0 kPa	23.0 °C

60.5 If gas at atmospheric pressure (101 kPa) and at 300 K is heated at constant volume to increase its pressure by 10%, what is the new temperature?

60.6 What is the percentage decrease in pressure when air at 15 °C is cooled to −5.0 °C at constant volume?

60.7 A rigid container risks rupturing if the pressure of the air within it rises above 230 kPa. It initially contains air at 110 kPa and 15 °C, and is sealed. What is the maximum temperature to which the air can be heated without risk of rupture?

60.8 Give the two reasons why the pressure of a gas goes up when it is heated in a closed container.

60.9 What is the special name for the temperature of −273 °C?

61   Charles' Law ♡

In this situation, the pressure is fixed (we use a container with a free-running piston). The gas is heated, and the volume increases.

As the temperature of the gas goes up, the average speed and kinetic energy of the molecules increases. This means that each time a molecule strikes the wall, its velocity change is larger, so the force on the wall is bigger.

However if the container expands, each molecule strikes the wall less often, leading to the same pressure as before.

The equation is

$$\frac{V_{after}}{T_{after}} = \frac{V_{before}}{T_{before}}$$

where T must be in kelvins.

Example – If I start with 30.0 cm^3 gas at $20.0\,°\text{C}$ and heat it up to $100\,°\text{C}$, what will the new volume be if I don't let the pressure build up?

1st stage: convert the temperatures to kelvins.

$$20.0\,°\text{C} + 273 = 293 \text{ K} \qquad 100\,°\text{C} + 273 = 373 \text{ K}$$

2nd stage: put the numbers into the equation.

$$\frac{V_{after}}{373 \text{ K}} = \frac{30.0 \text{ cm}^3}{293 \text{ K}}$$

3rd stage: rearrange the equation so that the thing you want to know is the subject, and calculate it.

$$V_{after} = 30.0 \text{ cm}^3 \times \frac{373}{293} = 38.2 \text{ cm}^3$$

4th stage: put the temperatures back in $°\text{C}$ if necessary (not needed here).

61.1 I start with 20 cm^3 gas at 30 °C and heat it to 200 °C. What will the new volume be if I don't let the pressure build up?

61.2 I start with 50 cm^3 gas at −20 °C and heat it until the volume is 100 cm^3 without letting the pressure build up. What will the new temperature be?

61.3 You want to store 150 litres of gas in a cylinder which only has room for 100 litres. You can do this by reducing the temperature. The 150 litres was measured at 15.0 °C. How cold will you have to make it in order that it will fit in the cylinder at the same pressure?

61.4 Work out the missing measurements from the following table, where each row is a separate question.

V_{before}	T_{before}	V_{after}	T_{after}
200 cm^3	300 K	(a)	600 K
200 cm^3	−23.0 °C	1000 cm^3	(b)
20.0 cm^3	(c)	200 cm^3	300 K
(d)	−183 °C	10.0 cm^3	23.0 °C

61.5 A gas thermometer is made of a narrow cylinder closed at one end, with a fixed mass of gas inside, and a tight-fitting yet low-friction piston at the other end. The piston moves to ensure that the contained gas is always at atmospheric pressure. The cylinder contains helium gas, occupying a length of 134.6 cm when at 22.4 °C.

(a) How long is the gas column when the temperature is −20.0 °C?

(b) How long is the gas column at −183 °C (oxygen boiling point)?

(c) How long is the gas column at 77 K (nitrogen boiling point)?

(d) How far will the piston move when the temperature changes by 1.00 °C?

(e) How far will the piston move if the temperature rises from 22.4 °C to 41.7 °C?

62   The General Gas Law ♡

Experiments have taught us…

Law	For fixed	In words	Formula
Boyle	Temp.	Halving volume doubles pressure	$pV = k_1$
Pressure	Vol.	Doubling temperature doubles pressure	$p = k_2 T$
Charles	Press.	Doubling temperature doubles volume	$V = k_3 T$

k_1, k_2 and k_3 are constants. The value of k_1, k_2, k_3 would depend on the control variable and the amount of gas in the experiment.

We must use the Kelvin scale so that zero (0 K) is the temperature of absolute zero. Temperature (K) = Temperature (°C) +273.

If you combine the three rules, you get

$$\text{pressure} \times \text{volume} = \text{constant} \times \text{temperature} \qquad pV = \text{const.} \times T$$

$$\Rightarrow \frac{\text{pressure} \times \text{volume}}{\text{temperature}} = \text{constant} \qquad \frac{pV}{T} = \text{const.}$$

Given that the constant must be the same before and after the process (as long as no gas leaks),

$$\frac{p_{\text{after}} V_{\text{after}}}{T_{\text{after}}} = \frac{p_{\text{before}} V_{\text{before}}}{T_{\text{before}}},$$

where T must be in kelvins.

Example – If I start with 10 cm^3 of gas at 20°C at a pressure of 101 kPa and heat it to 100 °C, what will the new pressure be if I let it expand to 12 cm^3?

1st stage: convert the temperatures to kelvins.

$$20\,^{\circ}\text{C} + 273 = 293 \text{ K} \qquad 100\,^{\circ}\text{C} + 273 = 373 \text{ K}$$

2nd stage: put the numbers into the equation.

$$\frac{p_{\text{after}} \times 12\ \text{cm}^3}{373\ \text{K}} = \frac{101\ \text{kPa} \times 10\ \text{cm}^3}{293\ \text{K}}$$

3rd stage: rearrange the equation so that the thing you want to know is the subject, and calculate it.

$$p_{\text{after}} = 101\ \text{kPa} \times \frac{10 \times 373}{12 \times 293} = 107\ \text{kPa}$$

4th stage: put the temperatures back in °C if necessary (not needed here).

62.1 I start with 5.0 cm^3 of gas at 30 °C at a pressure of 101 kPa and heat it to 200 °C. What will the new pressure be if I also compress it to 3.0 cm^3?

62.2 Work out the missing measurements from the following table.

p_{before}	V_{before}	T_{before}	p_{after}	V_{after}	T_{after}
101 kPa	200 cm^3	300 K	(a)	300 cm^3	600 K
101 kPa	150 cm^3	−23 °C	505 kPa	(b)	500 °C
10.1 kPa	24 cm^3	(c)	101 kPa	48 cm^3	300 K
5.0 kPa	(d)	−183 °C	50 kPa	2 000 cm^3	23 °C

62.3 I start with 24 L of air at room temperature, 20 °C, and pressure, 101 kPa. Calculate the new volume if:

(a) I raise the temperature to 303 K and compress it to 202 kPa;

(b) I raise the temperature to 600 K and let it expand to 80 kPa;

(c) I cool it to −120 °C and compress it to 303 kPa;

(d) I quarter its pressure and its Kelvin temperature;

(e) I heat it to 1100 K and compress it to 1.50 MPa.

62.4 The gas in a $0.020\,\mathrm{m}^3$ cylinder is stored at 80 bar $(1\,\mathrm{bar} = 10^5\,\mathrm{Pa})$. It starts at 290 K. The valve is opened until the gas pressure has equalized with the atmosphere (1.01 bar). Assume that all of the gas is now at 280 K. How much volume does it take up?

62.5 I begin cycling with the air in my tyres at 270 kPa and 285 K. Some time later, the air has warmed to 330 K, and the volume has increased by 3%. What is the new pressure?

62.6 A gas bubble of volume $3.0\,\mathrm{cm}^3$ forms at the bottom of a loch where the pressure is 3.0 atmospheres and the temperature 4.0 °C. What is its volume on reaching the surface where the water temperature is 13 °C?

62.7 The pressure in a flexible plastic flask is 1 000 kPa when its volume is $500\,\mathrm{cm}^3$ and its temperature is 10 °C. What would the pressure become if the gas volume was reduced to $400\,\mathrm{cm}^3$ and it was heated to a temperature of 90 °C?

62.8 A syringe contains $100\,\mathrm{cm}^3$ of a gas at 20 °C and its pressure is 1.0 atmosphere. Calculate the volume occupied by the gas if the pressure is increased to 1.5 atmospheres and the temperature becomes 240 °C.

62.9 A gas of volume $500\,\mathrm{cm}^3$ is initially at a pressure of 1.0 Atm and temperature of 17 °C. Its pressure is then increased to 1.5 Atm and its volume decreased to $400\,\mathrm{cm}^3$. What is the resulting temperature of the gas?

62.10 The gas in a spherical balloon is initially at 17 °C. The temperature of the gas increases so that the pressure increases by 2% and the radius of the balloon increases by 4%. What is the new temperature of the gas (in celsius)?